Understanding
TOXICOLOGY

Chemicals, Their Benefits and Risks

Understanding
TOXICOLOGY

Chemicals, Their Benefits and Risks

H. Bruno Schiefer, D.V.M., Ph.D.
Donald G. Irvine, Ph.D.
Shirley C. Buzik, B.S.P., M.Sc.

Toxicology Centre
University of Saskatchewan
Saskatoon, Saskatchewan, Canada

CRC Press
Boca Raton   New York

Publisher:	Robert B. Stern
Project Editor:	Renee Taub
Marketing Manager:	Susie Carlisle
Direct Marketing Manager:	Becky McEldowney
Cover design:	Dawn Boyd
PrePress:	Gary Bennett
Manufacturing:	Sheri Schwartz

Library of Congress Cataloging-in-Publication Data

Schiefer, H. Bruno, 1929–
    Understanding toxicology : chemicals, their benefits and risks /
  H. Bruno Schiefer, Donald G. Irvine, Shirley C. Buzik.
       p.   cm.
    Rev. ed. of: You and toxicology. 1986.
    Includes bibliographical references and index.
    ISBN 0-8493-2686-9 (alk. paper)
    1. Toxicology.   I. Irvine, Donald G.   II. Buzik, Shirley C.
III. Title
RA1211.S355   1997
615.9--dc20
                                    96-41143
                                        CIP

No claim to original U.S. Government works
International Standard Book Number 0-8493-2686-9
Library of Congress Card Number 96-41143
Printed in the United States of America 1 2 3 4 5 6 7 8 9 0
Printed on acid-free paper

PREFACE

This book, *Understanding Toxicology: Chemicals, Their Benefits and Risks*, grew out of an earlier one (1986) that was written for a limited readership. That book, *You and Toxicology*, was directed primarily to western Canadians, in response to inquiries concerning toxicological issues that poured into the Toxicology Research Centre at the University of Saskatchewan. The popular little book eventually found its way to CRC Press where the editorial group asked us to produce a similar guide to toxicology but for a larger audience.

The present book is written for the general reader — for everyone interested in chemical safety, toxic risks, pollution, hazardous wastes, and how to better understand and evaluate the often conflicting articles and news items about such topics. It is written by toxicologists, but not in technical language, and a glossary is provided for those technical terms that had to be used.

The authors are grateful for comments and critiques from colleagues, students, and members of the public. In particular, we appreciate the comments received from Dr. Michael Kamrin, Institute of Environmental Toxicology, Michigan State University, East Lansing, MI. Our gratitude goes to Dr. Michael Derelanko, Department of Toxicology, AlliedSignal, Inc., Morristown, NJ, for his comments and for bringing our earlier book to the attention of CRC Press. We thank Dr. K. Wayne Hindmarsh, Dean, Faculty of Pharmacy, University of Manitoba, Winnipeg, MB, for revising and updating the section on Street Drugs in Chapter 15, "Chemicals Misused."

Our thanks go to Carol Kettles for word processing the manuscript through its many drafts, and to Doug Hancock, both for producing the index and for his insightful comments and critical input.

H.B. Schiefer
D.G. Irvine
S.C. Buzik

DISCLAIMERS

Although every care has been taken in preparing *Understanding Toxicology: Chemicals, Their Benefits and Risks*, the authors accept no responsibility or liability for any consequences arising from information and opinions given in this book.

Names of commercial products in this book are printed with the first letter of the product capitalized and with a "registered"® symbol. The use of such trade names is merely meant to provide examples. No endorsement (or criticism) of any product is intended, by inclusion or exclusion, of any trade name.

THE AUTHORS

H. Bruno Schiefer, D.V.M., Ph.D., received a D.V.M. degree in 1956 and the equivalent of a Ph.D. degree in Veterinary Pathology in 1965, from the University of Munich. In 1966–1967, he was a visiting professor at the University of Connecticut, and in 1968 he was one of the international observers of the seal hunt in the Gulf of St. Lawrence. Dr. Schiefer joined the Department of Veterinary Pathology, University of Saskatchewan, in 1969. He was Acting Head (1971–1972) and Head (1974–1977) of that Department, followed by a sabbatical in Zurich, Switzerland (1977–1978).

From 1978 to 1983, he was Chairman of the Toxicology Group, College of Graduate Studies and Research, and worked towards the establishment of a Toxicology Research Centre at the University of Saskatchewan. Such a Centre was established in October 1982 and Dr. Schiefer was appointed Director, a position he maintained until the end of 1995.

His research interests were, at first, animal diseases in general, with particular emphasis on pulmonary pathology and diseases caused by fungi. For the last 25 years, he has been working on toxicology. He has produced over 320 scientific publications, and has spoken at many national and international conferences.

He was asked by the Department of External Affairs, Canada, to conduct a study on the alleged use of mycotoxins as chemical warfare agents in Southeast Asia. His report became an official UN Document in June 1982. Until 1993, he served as a consultant to the Arms Control and Disarmament Division of the Department of External Affairs.

Dr. Schiefer was instrumental in the establishment of the Canadian Network of Toxicology Centres and served as Director of its Western Node.

He was President of the Society of Toxicology of Canada from 1983 to 1985. In 1995, the Society honored Dr. Schiefer with the Award of Distinction "in recognition of exceptional contribution to his profession and the Society."

Donald G. Irvine, Ph.D., is Adjunct Professor of Toxicology in the College of Graduate Studies and Research at the University of Saskatchewan, and Co-principal Investigator in the Prairie Ecosystem Sustainability Study (PECOS), an interdisciplinary research program at the Universities of Saskatchewan and Regina.

Born and raised in western Canada, he obtained his B.A. degree and M.A. from the University of British Columbia and in 1981, a Ph.D. degree from the University of Saskatchewan.

As Research Scientist with the Saskatchewan Department of Public Health for 25 years, he identified metabolic abnormalities associated with neurological and psychiatric disorders, and investigated psychotoxic agents. Transferring in 1983 to the newly created Toxicology Research Centre at the University of Saskatchewan, Dr. Irvine extended his interests into general and environmental toxicology. He was national Team Leader of the Bioindicators Program in the Canadian Network of Toxicology Centres.

Dr. Irvine is a member of the Society of Toxicology of Canada, the Society of Environmental Toxicology and Chemistry (Europe), and the Canadian and Saskatchewan Societies of Clinical Chemists. He has written over 100 scientific papers, technical reports, and book chapters, and has presented lectures around the world. In the past 3 years, he has focused on the teaching and application of broadly interdisciplinary strategies for toxic risk assessment and for research into sustainability of ecosystems.

Shirley C. Buzik, B.S.P., M.Sc., was born, raised, and educated in Saskatchewan, Canada. She earned a Bachelor of Science in Pharmacy (1962) and a Master of Science in Toxicology (1986) from the University of Saskatchewan.

Prior to resuming her studies at the university, Ms. Buzik was dispensary supervisor and practicing pharmacist in a community pharmacy in Saskatoon.

From 1985 to 1996, Ms. Buzik worked at the Toxicology Research Centre, University of Saskatchewan, as Research Scientist responsible for public education and information services in toxicology. During this time, she wrote over 50 publications, including review articles, papers in refereed journals, and technical reports relevant to the field of toxicology. She was also responsible for responding to several hundred toxicology inquiries annually from the public, industry, government, and university departments in Canada and the U.S. In 1996, Ms. Buzik joined the International Consultants in Toxicology (ICT) Group, as President and Chief Executive Officer.

Ms. Buzik was national Team Leader of the Public Education and Communication Program in the Canadian Network of Toxicology Centres.

Ms. Buzik is a member of the Society of Toxicology (U.S.) and the Society of Toxicology of Canada. She has served as Councillor on the Board of Directors of the Society of Toxicology of Canada, and on the Board of Directors of the City of Saskatoon Community Health Unit.

TABLE OF CONTENTS

Chapter 1

WHAT IS TOXICOLOGY?

TOXICOLOGY DEFINED

The relatively young science of toxicology, or the science of poisons, is the study of the harmful effects of chemical and physical agents on living organisms. Scientists who study these harmful effects and assess the probability of their occurrence are called toxicologists. In this book, the word "chemical" will often be used in a very general sense; in such cases it can be understood to apply to physical agents as well. Chemical and physical agents that cause harm by virtue of their toxicity are also called toxic agents or toxicants.

WHAT IS TOXICOLOGY?

- Toxicology encompasses the study of the adverse effects of chemical and physical agents on living organisms and groups of organisms.
- Toxicology assesses the probability of hazards caused by such effects.
- Toxicology estimates the results of these effects on individuals, populations, and ecosystems (a complex of a community and its environment functioning as an ecological unit in nature).
- Toxicological studies consider the cause, circumstances, effects, and limits of safety of harmful effects of food, food additives, drugs, and household and industrial products or wastes.
- Toxicological studies deal with adverse effects ranging from acute to long-term.

HISTORY

"Practical toxicology," if we want to distinguish this from the "science of toxicology," is as old as humans themselves. In the early days of civilization, humans, in their search for food, must have attempted to eat a variety of plants and animals. Some of these were found to be safe and were used as food,

1

whereas others, found to be harmful, were rejected as "poisons." Today, we realize such a strict separation cannot be made. Rather, it is more reasonable to say there are degrees of harmfulness or safeness for any substance.

The many new chemicals introduced since World War II have brought with them an increased public awareness that chemicals may not only be beneficial, they may also be harmful. At that time, the science of toxicology, as we know it now, was born.

HOW MANY CHEMICALS HAVE BEEN DISCOVERED OR MADE?

November 1977: 4,000,000
May 1985: 7,000,000
October 1994: 13,000,000

- Vast majority isolated from natural materials or synthesized for research purposes.
- Most have been identified in the laboratory, but have found no use.
- Number of chemicals in common use: 60,000–70,000.
- Only about 5% of those in common use account for most of the weight of chemicals used.

SCOPE OF TOXICOLOGY

Toxicology has developed into four major but overlapping areas: environmental, economic, forensic, and clinical toxicology. **Environmental toxicology** deals with the **incidental** exposure of living systems to toxicants in the environment (food, water, air, or soil). Consuming food or water contaminated with either naturally occurring or synthetic (made by humans) toxicants or their residues is one example. Another example is chance exposure to foreign chemicals (xenobiotics) during occupational or recreational activities.

Economic toxicology deals with the harmful effects of chemicals **intentionally** administered to living things for the purpose of achieving a specific effect. The use of drugs to treat a disease (e.g., antibiotics to get rid of bacterial infection) or the use of pesticides to get rid of pests, such as mosquitoes, are two examples.

Forensic toxicology (relating to law) is concerned with the medical and legal aspects of adverse effects of toxicants on living systems. The medical aspects include the diagnosis and treatment of poisoning and are considered **clinical toxicology**, but the **legal** aspects require gathering information which relates to the cause–effect relationship between chemical exposure and the adverse effects. Both forensic and clinical toxicology use analytical methods to detect and quantitate the amount of a chemical in a living system. Both intentional and accidental conditions of exposure to toxicants are included in forensic and clinical toxicology.

There are a number of different "branches" of toxicology, as illustrated in the box. Although this is not a complete list, it indicates that toxicology is a diverse science.

SOME "BRANCHES" OF TOXICOLOGY

- Analytical toxicology
- Aquatic toxicology
- Clinical toxicology
- Ecotoxicology
- Environmental toxicology
- Epidemiological toxicology
- Forensic toxicology

- Immunotoxicology
- Nutritional toxicology
- Occupational toxicology
- Radiation toxicology
- Regulatory toxicology
- Toxicopathology
- Veterinary toxicology

THE TWO SIDES OF TOXICOLOGY

THE CHEMICAL ASPECT OF TOXICOLOGY

The word "chemical" has become a dreaded expression in modern society. We are warned constantly about the presence of chemicals in our food, water, air, as well as soil, and the harm they are doing to us and to the world in which we live. As a result, the word "chemical" produces visions of death, damage, and disease in the minds of many people. This, of course, is a poor view of chemicals. In fact, there are many benefits that chemicals offer society, be they drugs or household chemical products, etc.

A chemical is any specific substance composed of chemical elements, such as oxygen, hydrogen, carbon, or nitrogen. Therefore, everything in our world is composed of chemicals, from volcanoes to trees and people. Many people are more concerned about synthetic (made by humans) chemicals than about naturally occurring ones. A common misconception is that chemicals made by nature are good, and those made by humans are bad. This is simply not true. We must recognize and overcome this misconception before we can think and talk sensibly about chemical substances. Toxicologists recognize that nature is far more ingenious and prolific than humans could ever be in devising "toxic" chemicals. The distinction between synthetic and natural chemicals is mainly theoretical, although chemists have indeed succeeded in synthesizing many chemicals which do not exist in nature. However, our bodies do not recognize the origin of a chemical, be it provided by nature or from the chemical laboratory.

In addition to being worried about chemicals, we have also become concerned about physical agents that might be harmful to us. Radiation, either ionizing (e.g., X-rays) or nonionizing (e.g., electromagnetic fields), and ultraviolet (UV) radiation from the sun are examples of physical agents which have the potential to be harmful. Nonionizing and UV radiation will be briefly discussed in Chapter 14. On the other hand, the use of UV radiation for medical purposes or natural or artificial radioactive substances will not be discussed in this book.

THE BIOLOGICAL ASPECT OF TOXICOLOGY

Chemicals usually cause their effects by interacting with cells to change the way those cells function. Chemicals can cause damage to living organisms in many ways. They may be explosive or corrosive, or may cause irritation (redness, blistering, swelling, burning, or itching), or sensitization (allergic) reactions. Some chemicals can also cause harm because of their toxicity. These are called toxicants. The toxicity of a chemical refers to its ability to cause damage at some site in the body. Commonly, there is a remote action that may damage an organ system (such as the liver or kidney); disrupt a biochemical process (such as carrying oxygen by the blood cells); or disturb an enzyme system. This is in contrast to local toxicity and corrosiveness, which cause damage at the site of contact.

Living things depend upon special structures and a specialized set of chemical reactions, plus many systems to ensure that all processes operate in a harmonious and integrated way. Toxicants may harm structures or enzymes required for controlling normal mechanisms in the body.

Effects of toxicants may be immediate or delayed. The effects may occur in the exposed individual or in subsequent offspring. To evaluate toxicity, a variety of studies must be undertaken (see Chapter 2, Toxicity Testing).

FACTORS INFLUENCING TOXICITY

HOW MUCH, HOW OFTEN, HOW LONG

Many factors are responsible for the effects of chemicals on living things, whether these effects are good, bad, or indifferent. The single most important factor that determines harmfulness or safeness of a substance is the dose, i.e., the amount or **"how much"** chemical is taken up by a living system. Water is an example of this. You cannot live without water, yet if consumed in large quantities over a short period of time, it can be harmful. On the other hand, strychnine, the rat poison, in small enough amounts was once used as a medicine. The physician, Paracelsus (1493–1541), phrased this well when he noted, "All substances are poisons, there is none which is not a poison." In other words, all substances have the capability of being toxic, under some circumstances, and to some degree. The relationship between the dose of the chemical and the effects produced in the organism is called the **dose–response relationship**. The dose–response relationship means that, in general, as the dose increases the effect increases, and as the dose decreases the effect decreases. This is probably the most important concept in toxicology. Figure 1.1 shows a typical dose–response graph (curve).

In addition to **"how much"** (the dose), it is also important to consider for **"how long"** (the duration) the exposure lasts and **"how often"** (the frequency)

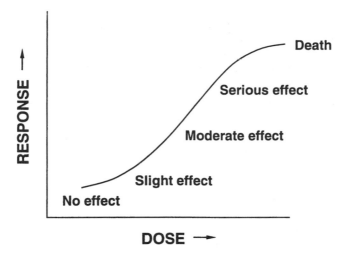

FIGURE 1.1 Dose–response relationship for a typical chemical. (From Kamrin, M. A., *Toxicology Primer,* 2.3, 13, Lewis Publishers, Boca Raton, Florida, 1988. With permission.)

the exposure occurs. This is known as the **dose–time relationship** and distinguishes two different types of toxicity. These are acute and chronic toxicity.

ACUTE AND CHRONIC TOXICITY OF CHEMICALS

The acute toxicity of a chemical is its ability to cause either local damage (e.g., to skin or eyes) or systemic damage (affecting the body as a whole), as a result of one exposure to a relatively large amount of that substance. This type of exposure is sudden and often produces an emergency situation, as when a child swallows a large number of acetylsalicylic acid tablets (ASA; Aspirin®*).

ACUTE TOXICITY VS. CHRONIC TOXICITY

- Acute effects (severe disease, death) do not predict chronic effects.
- Doses causing chronic effects (debilitating disease, cancer, malformations in offspring) may not cause acute or subacute effects.
- Some chemicals cause invisible damage to an organ or system (e.g., liver) which, in turn, is responsible for disease (e.g., hepatitis).

Chronic toxicity of a chemical is its ability to cause damage as a result of repeated exposure to relatively small amounts over a prolonged time period,

* Registered trademark in Canada of Bayer Inc., Etobicoke, ON.

such as repeated consumption of contaminated food or water. Chronic exposure usually does not produce effects until exposure has continued for some time.

Acute and chronic toxicities must be distinguished from each other. Acutely toxic effects are more readily apparent and more easily studied than are chronic effects. Acute and chronic exposures to the same chemical may produce quite different and unrelated symptoms. Acute toxicity cannot be predicted from chronic effects nor can chronic toxicity be predicted from acute effects. An example is arsenic toxicity. Acute arsenic toxicity causes mainly gastrointestinal tract symptoms such as vomiting and severe diarrhea, whereas chronic arsenic intoxication produces skin changes and damage to the liver, nerves, and blood-forming system.

WHAT IS THE LD_{50}?

A term commonly used to describe one type of acute toxicity is LD_{50}. LD means "lethal dose," and the subscript 50 means that dose of toxicant which is acutely lethal to 50% of test organisms under controlled laboratory conditions. The LD_{50} is a statistically derived value. The units of LD_{50} are usually milligrams of chemical per kilogram of body weight (mg/kg BW) of the organism. The *smaller* the LD_{50} (i.e., the fewer the milligrams of chemical per kilogram of body weight required to kill 50% of organisms), the *greater* the toxicity. Conversely, the larger the LD_{50}, the lower the toxicity. Table 1.1 depicts acute LD_{50} values for a variety of toxic agents. Some toxicants will cause death in microgram (1/1000 of a milligram) doses, whereas others may be relatively harmless following doses larger than several grams (1 gram [g] = 1000 mg). Toxicity of a substance thus can be rated from practically nontoxic to extremely toxic.

Every one of us ingests many lethal doses of synthetic and naturally occurring chemicals during our lifetime. There is a lethal dose of caffeine in approximately 100 cups of strong coffee and a lethal dose of acetylsalicylic acid (e.g., Aspirin®) in a bottle of tablets. The reason we survive so many lethal doses of so many chemicals, is that we do not drink 100 cups of coffee or take a bottle of pain tablets at one time. We take our poisons "in divided doses," not all at once. Our bodies are capable of handling smaller amounts of toxicants over a period of time. It takes the body time to destroy (metabolize) or get rid of (excrete) these substances. This shows the importance of the dose–time relationship (i.e., "how much," "how long," and "how often") in determining whether a substance will actually produce the toxic effects which it is capable of producing.

ROUTES OF EXPOSURE

In addition to dose, frequency, and duration of exposure, toxicity of a chemical is also dependent on the manner in which a substance enters the body. This is called the route of exposure. The three principal routes of exposure are: the gastrointestinal tract (oral), the skin (dermal), or the lungs (inhalation). Except for purely surface effects, if a toxicant cannot penetrate

TABLE 1.1
Approximate Oral (given by mouth)
Acute LD_{50} Values in Rodents for a
Selected Variety of Chemical Agents

Agent	LD_{50} (mg/kg)
Sodium chloride (table salt)	4,000
Ferrous sulfate (to treat anemia)	1,520
2,4-D (a weed killer)	368
DDT (an insecticide)	135
Caffeine (in coffee)	127
Nicotine (in tobacco)	24
Strychnine sulfate (still used to kill certain pests)	3
Botulinum toxin (in spoiled food)	0.00001

the skin, it will not be toxic by the dermal route of exposure; if it cannot become airborne, it will not cause toxicity by inhalation; and if a toxicant cannot be absorbed from the gastrointestinal tract, it will not cause toxicity by ingestion. There are few toxicants equally toxic by all three routes of exposure. Generally, a toxicant is most toxic by the route which permits fastest and greatest entry into the body. Some chemicals, notably certain pesticides, enter the body very readily through the scrotum.

OTHER FACTORS

Other factors that influence toxicity are species (type of animal), sex or gender (male or female), age, nutritional status, state of health, individual sensitivity, and the presence of other chemicals. Species differences (e.g., insects vs. humans) are responsible for observed variation in responses to some toxicants (e.g., pesticides). Frequently, these differences are the result of the way the body handles (i.e., absorbs, metabolizes, or excretes) a substance. Although there are species differences between humans and animals, there is frequently a sufficient degree of similarity, so that animal experiments are useful in estimating toxicity in humans.

Male and female differences in toxic responses have also been noted. These differences are likely due to different sex hormones. In laboratory animals, some chemicals show marked sex differences in acute and chronic toxicity. As examples, male rats are ten times more sensitive than female rats to the pesticide DDT, and only male rats develop cancer when exposed to certain hydrocarbons. In other cases, females may be more susceptible.

Age differences also contribute to some unexpected toxic responses. The very young or very old frequently react differently to toxicants. These differences may be caused by age-related changes in enzyme systems which can degrade chemicals, or in organ functions (liver, kidney) or in the body's defense mechanism (immune system), which is not as responsive in infants and the elderly.

Poor nutrition, poor health, and individual susceptibility frequently modify chemical responses. A poorly nourished, generally unhealthy human or animal is probably more susceptible to toxicants than a well-nourished, healthy one. Individual susceptibility to a chemical is well documented; some people react to just about everything, whereas others react to only a few substances.

Toxicity can also be influenced by the presence of other chemical substances. In some cases toxicity may be increased (synergism), in others decreased (antagonism). Synergism may be likened to $2 + 2 = 5$, whereas antagonism is like $2 + 2 = 3$.

Chapter 2

TOXICITY TESTING

TYPES OF TOXICOLOGICAL TESTS

Although the toxic hazard rating of a chemical (its potential to cause harm) is important, the critical factor is the risk associated with its use. Risk is the probability (likelihood) that a substance will produce harm under specific conditions of use. Safety is the probability that harm will not occur under specified conditions. Depending on the conditions of use, a very toxic agent can be less risky than a relatively nontoxic one. Toxicologists evaluate the nature of adverse chemical effects and assess the probability of their occurrence.

To evaluate toxicity, studies with different time frames are undertaken: acute, subacute, or chronic. Tests with mammalian species (such as rodents, rabbits, dogs, monkeys, etc.) are preferred, because the way other mammals deal with a chemical is more likely to be similar to the way humans do. In general, toxicants are most frequently tested in rats and mice. However, it must be realized that there is no one species of animals that handles all chemicals in exactly the same way as humans. Therefore, no single species can be used to estimate the toxicity of all toxicants to humans.

The type of testing required is determined by the nature of the risk involved. A food additive or a new drug, which is likely to be used by many people over a long period of time, will be subjected to intensive and prolonged animal studies and clinical trials. A new chemical which is to be used only sparingly and which would involve little human exposure, does not require as rigorous testing. An outline of some types of toxicological tests is provided in Table 2.1.

Toxicity testing experiments are usually done on all new chemicals, such as drugs, pesticides, food additives, household products, and industrial chemicals. Older chemicals are also sometimes retested. Animal research is usually done for the benefit of humans, e.g., prevention or treatment of human disease and suffering.

Several alternative non-animal testing methods (called *in vitro*, literally "in glass") are currently being developed in toxicology to reduce animal use or to replace animals entirely and still provide useful information. Some *in vitro* test methods use a variety of types of cells, organs, or tissue cultures. Lower organisms, such as bacteria, algae, fungi, plants, and insects, are also used. Using a computer, mathematical models based on the structure of a chemical (called the structure–activity relationship) have also been developed.

TABLE 2.1
An Outline of Some Types of Toxicological Tests

Acute toxicity testing — Single or repeated doses with observation for 14 days
  (typically tests for lethality or skin irritancy)
Subacute toxicity testing — Repeated doses for up to 90 days
Chronic toxicity testing — Repeated doses for up to 2 years (includes carcinogenicity testing)
Special tests

- Teratogenicity, reproductive, and developmental toxicity testing
- Mutagenicity and genotoxicity testing
- Toxicokinetic and metabolic studies
- Skin and eye effects
- Behavioral studies
- Immune system studies

The major problem with these alternative methods is validation. Validation seeks to answer the question "How do results from these *in vitro* test methods compare with those from animal studies?" The use of these methods cannot fully simulate the effects of a toxicant in a complex living system. It is very difficult to find a substitute for a human tissue or organ, much less for the whole body. Because *in vitro* methods do not take into account the effects of the circulation of the blood or the actions of the nervous system, none of the alternative methods (as they exist today) will entirely eliminate the need for animal testing. However, toxicologists are working hard to develop improved alternative methods and groups of methods which eventually can be validated. Unfortunately, the process requires a great deal of time, effort, and money.

Conventional toxicological testing is also expensive, and the usual battery of tests to evaluate a new chemical may cost several million dollars and take years to complete.

ACUTE TOXICITY TESTING

Acute toxicity tests involve giving a chemical, on one or more occasions (within a 24-hour period), to determine some toxic outcome. The animals are closely observed for 14 days, and all adverse effects and deaths are recorded during this time. To assess lethal effects, statistical methods are used to calculate the LD_{50}. As mentioned previously, the LD_{50} value is usually expressed as milligrams per kilogram of body weight (mg/kg BW). A comparison of LD_{50} values shows wide variation among species, for some chemicals. More recently, newer acute toxicity test methods (e.g., Fixed-Dose Procedure; "Up-and-Down" Study) have decreased the number of animals used in these experiments.

SUBACUTE AND CHRONIC TOXICITY TESTING

In subacute testing, toxic effects are studied in an animal population exposed daily to a toxicant for a tenth of the animal's lifetime. As an example, in rats this is about 3 months. The subacute study examines adverse effects other than death. These may include adverse effects on behavior, organs, or body fluid components, among others.

In contrast to the acute and subacute tests, chronic (long-term) toxicity tests are those in which the chemical is administered for a substantial portion of the lifetime of the test animal. The purpose of chronic toxicity tests is to obtain information about the adverse effects of exposure to a relatively small amount of a chemical for a prolonged period of time. A 2-year study in rats resembles human exposure lasting 68 years, whereas a 2-year study conducted with monkeys is equivalent to only 8.9 years of a human life.

Special tests for effects on the immune system, skin, eyes, and behavior may also be needed in addition to reproductive and developmental effects, including teratogenicity (causing malformations), carcinogenicity (causing cancer), and mutagenicity (affecting the genes) studies. Other tests such as toxicokinetic studies, including metabolism, may be included as required. Toxicokinetic studies (called "pharmacokinetic" when referring to drugs) examine how the body handles toxicants. Such studies determine whether the substance accumulates in a specific body organ or tissue, and how the body transports, transforms (metabolizes), and excretes the substance. Metabolism usually decreases the toxicity of a chemical, but in some cases it increases toxicity instead. In most animal species, the liver is the primary organ of metabolism.

REPRODUCTIVE AND DEVELOPMENTAL TOXICITY TESTING

Reproductive toxicology is the study of the adverse effects of toxicants on the reproductive system. It was once thought that only the female system was affected. However, now we know that male reproduction can also be harmed.

Reproduction is a very intricate, complicated process in both sexes. It is regulated by the endocrine system (a complex network of glands that secrete hormones). Any toxicant that disrupts endocrine function (called endocrine disruptors) can also interfere with reproduction. Exposure to such agents (e.g., DDT) has been linked to decreased fertility, and demasculinization, and defeminization in several species.

Certain drugs (e.g., diethylstilbestrol, some cancer treatment agents), pesticides (e.g., chlordecone and lindane), industrial chemicals (e.g., *n*-hexane, ethylene glycol), radiation (e.g., X-rays), metals (e.g., lead, methylmercury) as well as alcohol, nicotine, marijuana, and cocaine, among others, affect the reproductive process in both sexes. Sperm abnormalities, decreased birth weights, stillbirths, decreased fertility, and infant deaths have been reported from such exposures.

Reproductive toxicity studies are of three basic types: those which study the effect of a chemical on the pregnant female; those in which a chemical is given to both males and females prior to mating and continuing until the young are weaned; and those that continue until three generations are produced. Recently, more tests have been developed to study the type and exact mechanism or location of toxic effects in male and female reproductive processes.

Developmental toxicology is the study of exposure of both sexes to toxicants thought to cause adverse effects in the developing organism (from conception to adolescence). Originally, only female exposure was considered damaging to the offspring, but recent reports also implicate the male. It is important to realize, however, that the majority of developmental defects are not caused by toxicants but by as yet unknown causes.

Developmental toxicity testing includes studies of structural malformations or birth defects (teratogenicity), growth retardation, functional impairment (e.g., neurobehavioral effects), and death of the offspring. Agents that are toxic to the fetus are called "fetotoxic" and those that are toxic to the embryo, "embryotoxic." So far, a few developmental toxicants have been identified in males (e.g., marijuana, opiates, lead, cigarette smoke, urethane, and the drugs diethylstilbestrol and cyclophosphamide). It is thought that such agents can damage the sperm directly or be present in the semen and could interfere with fertilization and early developmental events.

The developmental defects resulting from female exposure have been known longer, and agents such as cigarette smoke, ethanol, cocaine, lead, organic mercury, and numerous drugs (e.g., retinoids, thalidomide, anticancer drugs) have been implicated.

One type of developmental toxicity is teratogenicity. Teratogenicity studies determine the ability of a substance to cause structural malformations or birth defects in the developing organism. A teratogen is a substance that causes such defects. Alcoholic beverages and cocaine are examples of chemical teratogens. It is important to realize that teratogens can cause malformations in the developing organism by direct effect on the embryo or fetus or indirectly through toxicity to the mother or the placenta, or by a combination of direct and indirect effects, and must do so during a critical phase in the gestation period when organ systems are being formed. The critical period in organ development in humans is the first 3 months (first trimester) of pregnancy. The wisest course for any man or woman considering conception is to limit their exposure to all toxicants as much as possible.

Females differ from males when it comes to potential teratological effects. Females are born with numerous eggs already in the ovaries. Thus, chemical residues that might accumulate in the female body could have negative effects in these cells. In contrast, the sperm cells of males are newly generated throughout life, and so, unlike eggs, cannot accumulate toxicants over several years.

If fertilization occurs and a fetus develops in the womb, it is the mother that supplies all the nutrients — but also the toxicants if they are present. The main way to prevent malformation of babies due to toxicants is a life-time awareness of the female's special role. One well-known example was the exposure of pregnant women to the drug thalidomide, which resulted in severe fetal malformations. Another area of concern may be infants who are breast-fed by mothers exposed to toxicants. If the toxicants are fat soluble they are likely to accumulate in the milk. However, the benefits of breast feeding may well outweigh the toxic risk.

GENOTOXICITY AND MUTAGENICITY TESTING

There are several different testing methods to determine genotoxicity, the ability of a toxicant to alter heredity. Some look for changes in the chromosomes (structures bearing the genes), and others measure the ability of a substance to change the DNA (the basis of the genetic code). A substance altering the structure of a gene is called a "mutagen." Mutagenicity tests utilize a variety of test organisms, both animal and non-animal. The best known of these tests is the Ames test, one of several short-term, *in vitro* tests. The Ames test is done with bacteria and can rapidly demonstrate whether a substance is a mutagen.

CARCINOGENICITY TESTING

Some of the same methods used to test for mutagenicity are used as screening tests for carcinogenicity, because mutation is usually a critical step in causing cancer. However, not all substances that are carcinogenic (cause cancer) are also mutagenic (cause mutation), and not all mutagens are carcinogens. Therefore, it is still necessary to undertake long-term tests for carcinogenicity. Because of the long time required for most cancers to appear and the low frequencies of cancers associated with low concentrations of carcinogens, it is necessary to use large doses of the toxicant in order to increase the possibility of detecting a carcinogenic effect in a limited number of animals.

Although it seems we are constantly being warned about all the substances that can cause cancer, in actual fact, there are very few substances that have been clearly proven to be carcinogenic in humans.

NUMBERS IN TOXICOLOGY

One ppb equals 0.001 ppm; what does that mean? The preferred method of expressing concentrations of toxicants in food, water, air, or tissues is to use parts per million (ppm) or parts per billion (ppb). Here are some mathematical considerations:

- 1 ppm equals one part in 1,000,000 parts, on a weight-by-weight basis; e.g., 1 ppm is 1 mg/kg or 1 μg/g or 0.0001%.
- 1 ppb equals 0.001 ppm.

Because it is so difficult to comprehend how small an amount 1 ppm is, the following are some approximations to illustrate this point:

1 ppm = 1 cm in 10 kilometers (1 inch in 16 miles)
= 1 minute in 2 years
= 1 cent in $10,000
= 1 large mouthful of food compared with a person's lifetime of eating
= 1 orange in one box-car full of oranges.

Consequently, 1 ppb equals 1 orange in 1,000 box-cars full of oranges.

No matter how small an amount 1 ppm or 1 ppb of a chemical appears to be, it is well known that such amounts, or even much smaller ones, can determine whether we live or die. While this may be difficult to grasp, it is a fact.

ANALYSIS FOR TOXICANTS: EXPENSIVE AND TIME-CONSUMING

Concentrations of toxicants in various commodities can be described in terms of parts per million or parts per billion. In fact, some toxicants are measured at the parts per trillion (ppt) level. Who comes up with such figures, and how are they determined?

The analytical toxicologist, or an analytical chemist, determines such small concentrations by using highly sophisticated techniques, such as thin layer chromatography, high performance liquid chromatography, gas chromatography, or gas chromatography coupled with mass spectrometry (GC/MS). In GC/MS, each molecule is broken apart at its weak points, and each individual fragment is measured. A computer, connected to the GC/MS, counts all the various fragments and displays the results on a monitor. Such an investigation does not take much time once it gets to the machine. However, before such a highly specialized analysis is done, the sample has to be prepared very carefully. This is called "extraction and clean-up," and is very time-consuming and labor-intensive. If the sample is not suitably prepared, the machine will produce invalid results: garbage in, garbage out!

Sometimes on television, viewers are fascinated by the speed with which toxicologists appear to come up with an answer. Unfortunately, in real life, this is not so easy. An analysis is very time-consuming and very costly (one GC/MS instrument alone can cost up to $1,000,000).

Analytical laboratories are often asked to investigate a sample (whatever it may be) for toxicants. As we have seen earlier, all substances can be toxicants. So, where does the analyst start to look for "all possible toxicants?" It would take a whole laboratory more than a year to apply all sorts of techniques and to analyze for all sorts of things. Even the richest person or company in the world would not be able to afford to analyze more than one sample. It is, therefore, very important to specify as clearly, and to define as narrowly as possible, what particular toxicant should be looked for. Even then, the costs of one analysis might be substantial.

Chapter 3

THE NEVER-ENDING DREAM: ZERO RISK

Figure 3.1 From Malcolm Hancock. With permission.

UNDERSTANDING HAZARDS AND RISKS

Life is fraught with numerous risks. Creation of a risk-free life is not even a dream; it is, plainly speaking, utopian. Many of the hazardous situations into which we place ourselves are of our own choice and under our own control — they are voluntary risks. We drive automobiles, knowing perfectly well that there is a risk of becoming involved in an accident. We consume stimulating beverages such as tea and coffee or stronger, intoxicating beverages (like alcohol), knowing that this will influence our reactions and behavior, and yet, we continue to drink them. The list can go on almost forever. Other risks to our health, however, are not under our control — they are involuntary risks. Those of us living or shopping in polluted city air, for example, may know that this is not particularly healthful, but we think we have to put up with it without complaining.

Indeed, humans have peculiar ways of perceiving risks. Publicity surrounding a risk plays an important role in shaping our views. The death of 500 people in an aircraft crash is certainly newsworthy, and demonstrates the risks of flying, but the same number of accidental traffic deaths of people, reported one by one on the back pages of the newspapers, over one week's time, does not seem as tragic. We willingly accept the risk of unlawfully crossing a busy highway (a voluntary risk), but we are quite upset about "stinking" diesel exhaust fumes (an involuntary risk), and assume that our health must be endangered.

To understand hazards and risks, it is first necessary to distinguish a "risk" from a "hazard." In toxicology, a hazard is the potential harm that a chemical can do, i.e., anything that presents a danger. Chemicals are, to varying degrees, hazardous because they have the potential to cause adverse biological (toxic) effects. Risk is the probability (likelihood) that harm will occur under specific circumstances.

HAZARDS AND RISKS OF THE NATURAL ENVIRONMENT

The natural environment provides many hazards and risks to health. Not all examples given here refer to chemicals, directly, but they are typical examples of natural risk:

- *Radiation.* It is well known that levels of cosmic rays increase with altitude. People flying in aircraft are exposed to higher doses of cosmic rays than people living just above sea level. Even if we do not fly, radiation emanating naturally from the Earth's crust affects us, too. At the same time, uptake of radionuclides (substances which emit radiation) from food and water unavoidably exposes humans to some radiation effects from such natural sources.
- *Minerals and Metals.* Many carcinogenic and toxic metals and minerals are found naturally in the crust of the Earth. While it is true that most of these compounds are liberated by mining and industrial activities, some may get into the food chain quite naturally.
- *Volcanoes.* The eruption of a volcano disperses incredibly large amounts of particles into the air. While many of the compounds released are either harmless or just a nuisance, depending on the viewpoint, some have been found to be carcinogenic.
- *Oxygen.* Air, containing oxygen, is needed to breathe and live, yet oxygen can be quite poisonous and can cause severe toxic effects. For instance, it can occasionally produce biologically dangerous "superoxide radicals" and peroxides in fat. It would be difficult to explain fully here, but superoxide radicals are known to affect the normal functions of mammalian cells and can destroy them.

- *Bacterial Products.* Various bacteria oxidize organic nitrogen and ammonia into a variety of nitrogen compounds, including N-nitroso compounds, known to be highly potent cancer-causing agents. Consuming food or water contaminated with toxins produced by microorganisms, such as bacteria, is one way to be exposed to a naturally occurring hazard. Botulism is an example of a disease caused by eating food contaminated with a bacterium (in this case the organism, *Clostridium botulinum*). This bacterium is of particular concern because the toxin produced is generally considered to be the most acutely toxic chemical known.

- *Fungal Products.* Antibiotics, derived from fungi, are very helpful drugs that save millions of lives, and yet, many secondary products of fungi are known to be extremely toxic. The health effects of fungi vary. Aside from causing deterioration of fabrics and building materials, fungi can produce allergic reactions in the respiratory system and on the skin of sensitive people. However, some fungi produce more serious health effects due to the production of toxins which are carried on spores or dust particles through the air. Aflatoxin, produced when *Aspergillus flavus* fungi grow in or on peanuts, for example, is one of the most potent cancer-causing agents we know. Other fungi produce toxins that can damage the immune system of the body or irritate the skin or the respiratory tract. Eating food or breathing air that contains only traces of toxic fungi (also known under the name of mycotoxins) seldom produces immediate or dramatic reactions. Vague symptoms of "not feeling well" may be the only signs, and the connection of the disease with mycotoxins can be easily overlooked. It is noteworthy that it is nature, not the laboratory, that provides these toxins.

- *Plants.* The toxic properties of many plants have been recognized for thousands of years. Humans have used such plants to treat diseases by either using whole plants or their extracts, as drugs. People have recognized that the use of certain plants (e.g., as herbal teas) induced pleasurable feelings, or thought that using them might provide pleasure (e.g., tobacco). To list just a selection of the numerous toxic plants would fill a book double the size of this one, but some examples will be mentioned in Chapter 8. Cabbage, broccoli, and turnips contain substances that can cause enlargement of the thyroid gland (goiter). Yet such vegetables also contain substances (e.g., vitamins, fiber) that are beneficial. When bees produce honey from tansy ragwort, a commonly occurring weed throughout the world, it can contain the poisonous substances known as pyrrolizidine alkaloids. These alkaloids are toxic to the liver and, eventually, to the lungs. Bracken fern is used as a salad ingredient in some countries. This fern has been linked to cancers of the esophagus, the urinary bladder, and the large bowel. Some plants, including soybeans and alfalfa, contain female hormones (estrogens). Estrogens from plant sources are called phytoestrogens. Although phytoestrogens are capable of causing infertility in animals grazing heavily on such fodder, the plants are not usually associated with such problems in humans.

- *Cooking.*   The use of fire in the preparation of food is seen as one of humanity's greatest achievements, because it allows us to eat a wide variety of food which would have been unpalatable otherwise, not very well tolerated or accepted by the body, or even toxic, when eaten raw. Nevertheless, almost every type of food preparation, such as salting, smoking, fermenting, barbecuing, roasting, or grilling, is capable of producing cancer-causing agents.
- *Air.*   We all object to air pollution from automobiles or factories, but it has to be recognized that there are many natural pollutants in the air. Forests, for instance, normally emit large quantities of terpene hydro-carbons (reported to be cancer causing) from the foliage into the air. Lightning, or ionizing cosmic radiation, produces nitric oxide in the air, and a forest fire emits large quantities of nitric oxide, nitrosamines, benzpyrenes, and many other potentially cancer-causing chemicals.

In summary, a review of these "hazards and risks of the natural environment" indicates that there is nothing in nature which we can eat, breathe, or drink without running some risk of being poisoned or encountering carcinogens. Thus, a picture of mind-boggling complexity evolves: Unless humans have developed methods in their bodies to cope with this onslaught of toxins and carcinogens, we are all doomed by nature to become extinct. The history of the human race tells us that the latter is, evidently, not the case. Life which is, indeed, fraught with risk, continues.

HAZARDS AND RISKS OF THE CHEMICAL ENVIRONMENT

Ever since industrial activities started to reshape the landscape — witness the ugly scars produced from the tailings of mining operations or the death (in past years) of almost all vegetation around the smokestacks of smelters — people have suspected that there was something "bad" and "dangerous" about such activities. However, all such signs of destruction of our environment were accepted as unavoidable; as local events that would not endanger planet Earth in general.

It was Rachel Carson's book, *Silent Spring*, which ushered in a new era in 1962. Carson warned that the widespread use of insecticides would eventually destroy the biosphere, and humans with it. The chief damage produced by chemicals, Carson said, was being done by synthetic or "unnatural" ones made by humans. For the first time since people inhabited the Earth, she said, they were exposed to hazardous chemicals from the womb to the grave. In other words, the ecological movement was born.

Whatever short-comings one may find with Carson's book, she was certainly right about her concerns over the widely used insecticide, DDT. Contrary to industry predictions, DDT showed up in the most unlikely places. For instance, the insecticide was found in the Antarctic, where it had never been used, or in food destined for human consumption, and eventually in the bodies of humans. The battle was on. But Carson never called for banning of all toxic agents. She asked for chemicals which would destroy the unwanted pests, but would not endanger humans.

QUANTIFYING RISK

We should not underestimate the risk of living with chemicals. Instead, we must remember there are no safe chemicals, but there are safe ways of manufacturing, handling, and using them. In fact, chemical safety does not mean the total absence of risk, but risk management of chemicals.

THE QUEST FOR CHEMICAL SAFETY

"But we should not underestimate the risk of living with chemicals. We all know there are no safe chemicals, only safe ways of manufacturing, handling, and using them."*

"Chemical safety does not mean the absence of any risk whatsoever; rather, chemical safety has to do with risk management."*

* From: *International Register of Potentially Toxic Chemicals Bulletin 7.* United Nations Programme, Geneva, Switzerland, 1985.

When talking about chemicals and their risks, we first have to consider a few points: who is at risk; how great is the risk; and what risk is posed by traces of chemicals? At highest risk are those employed in manufacturing chemicals. High rates of disease occurring in workers manufacturing certain chemicals indicate that such chemicals are toxic. At second highest risk are people who use dangerous chemicals in large quantities. The lowest risk is experienced by those who purchase chemically contaminated food or commodities.

To a great extent, advances in technology have helped, or could help, to avoid a wholesale contamination of the environment, or to protect workers, applicators, and users. Curiously, progress in analytical technology for the detection of very minute quantities of chemicals has outpaced the capacity to understand the health impacts of such small quantities. Trace amounts of chemicals can now be detected in drinking water, and the health problems of a population exposed to such water can be tabulated, but there is a gap. How can we be sure that headaches, sleeplessness, a feeling of "being down", and similar nonspecific symptoms are not actually caused by these trace amounts?

To define the health risk from chemicals in a major industrial catastrophe is relatively easy. But to define that risk from chemical pollution by trace amounts of toxicants is quite another matter. There are a few proven examples that trace amounts can cause disease. An example is diethylstilbestrol (DES), once used as a growth-promoting hormone for beef cattle. The same drug, taken by women to prevent miscarriages during pregnancy, caused cancer in the reproductive tract in female offspring and physical abnormalities (noncancerous) in male offspring. From such an example, it must be concluded that it is prudent to restrict the use of chemicals to the absolute minimum, and to continue to demand that all possible testing procedures are performed and evaluated before a chemical is released into the environment.

RISK ASSESSMENT

Risk assessment is a process which evaluates the probability that harm will occur under specified conditions. In this process, the relevant biological, dose–response, and exposure information (data) are combined to produce an estimate of the probability of adverse outcomes from exposure to a specific chemical agent or mixture.

A major approach to assessing risk is Quantitative Risk Assessment (QRA). This approach uses two types of information. One type is information that is already available (e.g., experimental laboratory information and epidemiological data). The second type is specific to the site, or concern, and has to be measured or predicted from technical models (e.g., exposure levels). Together, these two types of information are used to estimate the probability of harm occurring to the exposed population.

There are four steps in risk assessment: (1) hazard identification (does the substance cause the specific harmful effect?); (2) dose–response assessment (what is the relationship between dose and incidence of adverse effects in humans or species of concern?); (3) exposure assessment (what exposures are currently experienced or anticipated under different conditions?); and (4) risk characterization (what is the estimated incidence of the adverse effect that will occur in the population, given the exposure levels estimated, under the prevailing conditions?). The information collected in steps 1 through 3 is used together in this fourth step (risk characterization) to provide a quantitative or numerical conclusion.

TABLE 3.1
Some Important Terms in
Quantitative Risk Assessment

- No Observed Adverse Effect Level = NOAEL
- Lowest Observed Adverse Effect Level = LOAEL
- Reference Dose = RfD
- Acceptable Daily Intake = ADI
- Uncertainty Factor = UF
- Modifying Factor = MF
- Slope Factor
- Air Unit Risk
- Drinking Water Unit Risk

As depicted in Table 3.1, the highest dose level that does not produce an adverse response is called the "No Observed Adverse Effect Level" (NOAEL). The lowest dose level that actually does produce an adverse response is called the "Lowest Observed Adverse Effect Level" (LOAEL). The Environmental Protection Agency (EPA) uses the NOAEL to calculate the Reference Dose (RfD). The World Health Organization (WHO) uses the NOAEL to calculate the Acceptable Daily Intake (ADI) for pesticides and food additives. In such calculations, there are Uncertainty Factors (UFs), such as a tenfold UF used in extrapolation from animal data to human data; a tenfold UF used to extrapolate from subacute to chronic exposure, etc. A Modifying Factor (MF) is used to account for uncertainties not otherwise addressed. Using all these factors, the EPA calculates the RfD and WHO calculates the ADI from:

$$\frac{\text{NOAEL}}{\text{UFs (multiplied together)} \times \text{MF}}$$

From this, it would seem that the RfD and the ADI will be identical for any selected toxicant, but this is not necessarily so, since different regulatory groups, working with different populations and often at different times, may assign somewhat different values to one or more of the factors, or may select their NOAEL based on a different toxic effect.

So far, this discussion of risk assessment has dealt with all situations in which there is a threshold dose or concentration below which the toxic substance has no harmful effect. Most chemicals fall into this category, and toxic risks from them can be estimated correctly using the methods just described. On the other hand, toxicants that cause cancer or which produce genetic mutations (carcinogens and mutagens) are generally considered to have no threshold, and for this reason, such risk assessment has to be altered. The four basic steps of risk assessment are still essential, but the NOAEL (which is an estimate of threshold) is not used, and the dose–response relationship has a different form. Three tools commonly used to estimate carcinogenic or mutagenic effect are the Slope Factor and the Unit Risk Factors for air and for water. These factors all express the cancer risk per unit dose, over a lifetime of exposure. Slope Factor (risk per mg/kg body weight per day) represents risk from dietary intake; the Drinking Water Unit Risk (risk per microgram/liter of drinking water), and the Air Unit Risk (risk per microgram/cubic meter of air), express the risk specifically through those media.

All of the risk-related measures discussed so far belong to what was called the first type of information — measures of hazard already available as published values from laboratory or epidemiological studies. They are generic or universal values. The second type of information essential for risk assessment is specific to the site, or concern, and has to be measured or predicted from mathematical models for the area involved. For a specific situation, the risk assessment must characterize the source(s) of the toxic materials, identify their pathways to the exposed population, estimate the population's exposure, and describe the population's susceptibility. Only a sample of the sorts of data often needed can be given here: amounts of the chemical at source; soil permeability and chemical access to groundwater; use of water wells in the area; distance to closest house; unusual food preferences in the area; exposure through local produce; presence of endangered species; exposure to "drinking water" contaminants through showering; preexisting medical conditions in the area; age-specific exposures (e.g., infant foods or consumption of dust and soil); and age-specific susceptibilities (e.g., fetal development, infants' immature immune systems).

Together, the two types of information described are used to estimate the probability of harm occurring to the exposed population. The RfD and the ADI are both measures of a threshold intake of a toxicant, below which it is highly unlikely that even a lifelong exposure will have any adverse effect. To use these measures, the expected daily intake of the toxicant from all sources, through all exposure routes, must be calculated and then compared to the RfD or ADI. For nonthreshold (mainly carcinogenic) effects, the procedure is different: predicted intakes are multiplied by the Slope Factor, or by the Air or Drinking Water Unit Risk values, to yield the lifetime cancer risk directly. That risk is normally expressed as probable cases of cancer per million people; most groups surveyed seem to agree that a less than one in 1,000,000 probability can be accepted, and some accept one in 100,000. In all quantitative toxic risk assessment, it is usually the incremental risk that is determined (i.e., how much will the risk rise?) rather than relative risk (how bad is this particular risk, compared with a different but relevant one?) or cumulative risk (how high will the total risk for this effect in the area become?).

Risk assessment is a useful tool to estimate probable harm, but it does have some limitations and uncertainties. These uncertainties arise because there are always gaps in our knowledge or a lack of understanding of how a substance causes its adverse effects. These gaps in knowledge are partially overcome by using extrapolations, mathematical models, or assumptions. Quantitative risk assessment is still undergoing development and refinement, but description of emerging methods is beyond the scope of this book. More information on risk assessment procedures can be found in the publications by Kamrin (1988), and by Kamrin, Katz, and Walter (1995), identified in the "Further Suggested Reading" section.

WEIGHING BENEFITS AGAINST RISKS

As explained above, it is extremely difficult to quantify a risk and to describe it accurately. Even when all the details of the probability of environmental or other risks are known, there is still the question of weighing risks against benefits. For example, which one is of greater benefit to society — not using a chemical and experiencing a possible loss in crop production, or using the chemical and possibly harming human health? To exaggerate a bit: if you have a choice of either dying from starvation or from chronic poisoning, which one do you prefer? Perhaps, with better understanding and use of toxicology, there is a third choice: careful use of selected chemicals and hence neither starvation nor chronic poisoning.

The process of trying to find a balanced answer to this question is known as risk–benefit assessment. It should not come as a surprise to learn that risk–benefit assessment is, to a large extent, a subjective interpretation of scarce data. Scientists and economists act with the best intentions when they calculate risks and benefits. In the end, it is the politicians who make the final decision, and politicians will be influenced by public opinion, informed or not.

PERCEPTION OF RISK

The risk of death through natural disasters (floods, hurricanes, earthquakes, and similar so-called "acts of God") is about one in 1,000,000. The risk of death of one in a million per year is of little concern to the average person, but risk of death of one in 100,000 per year causes people to give warnings (e.g., parents may tell children not to go into deep water or participate in hang-gliding activities). When the risk rises to one in 10,000, the public is prepared to pay for its reduction (e.g., acceptance of laws requiring the use of safety belts when using automobiles).

All attitudes can change, or are relative to expected benefits. Perception of risk is influenced by a person's estimate of the benefits (monetary or otherwise) he or she will get from taking the risk. The greater the expected benefits (real or imagined), the more risk a person is willing to accept. As far as toxicology is concerned, the better informed the public is about toxicological, chemical, and environmental issues, the firmer and more valuable will be the expression of their opinions.

THE RIGHT TO KNOW

The right of the consumer to be advised of the hazards of product misuse and to be provided with information on safe handling practices (and emergency and first-aid procedures) has long been recognized in law. Numerous statutes and regulations concerning the packaging, labeling, transport, storage, and use of hazardous products are in place in most developed countries.

It is now accepted that there is also a need to provide information on chemical products which are being used by workers (during manufacturing, or in laboratories, etc.) and to users of large quantities of chemicals.

It is not just as users of products in the home or in the workplace that people may become exposed to harmful substances. We have been alarmed by the chemical disaster in Bhopal, India (1984) and the nuclear disaster in Chernobyl, Ukraine (1986). Media coverage of other, less tragic but nevertheless serious incidents have also alarmed us, such as PCB (polychlorinated biphenyls) spills from transformers being transported across the country, or train derailments where tanker cars containing large quantities of highly dangerous chemicals were damaged and released their contents. No wonder people are worried.

People who are being put at risk because of such events should be provided with the information necessary to assess the adequacy of control measures, monitoring activities, and emergency response plans that are in place. The public has a "right" to know the following types of information:

- The sort of hazard that might exist
- An indication of the degree of risk
- The specific nature of the effects and their likelihood
- Information on relevant protective measures
- Legal requirements relating to the hazard
- Action to be taken in the event of an accident

Disclosure of information on any of these points goes a long way toward obtaining the full cooperation of the public, and toward making all of us willing and helpful partners.

Chapter **4**

PESTICIDES

The definition of a pest is a living organism which is not wanted; it usually refers to a whole species or group of organisms, present in considerable quantity, and at a time and in a place where the efficiency of some production or activity is threatened.

Pesticides, then, are agents (usually chemicals) intended to kill (or severely disable) unwanted species of animals or plants. This large group of chemicals can be classified most meaningfully by the intended target of their damaging action: insecticides (e.g., malathion, carbofuran) are meant for killing insects; herbicides (e.g., 2,4-D, picloram) are meant for killing herbs, or more generally, for killing plants; rodenticides (e.g., warfarin) are meant for killing rodents; piscicides (e.g., rotenone) are used to kill fish; algicides (e.g., copper sulfate) are used to kill algae; and fungicides (e.g., benomyl) are used to kill fungi.

There are other ways of getting pests out of an area, besides killing them with chemicals. One is to use chemicals that prevent pests from mating, reproducing, or from developing properly. Another is to lure them into traps, using natural attractant chemicals, and still another is to selectively infect them with specific diseases or parasites. Sometimes natural or introduced predators are used to feed on them. Gulls are famous for their ability to eat enormous numbers of grasshoppers, and in Salt Lake City, Utah, there is a monument to them, for their role in eradicating a plague of "hoppers." Another approach is to farm in ways that reduce the numbers of pests to a minimum. Then, smaller amounts of chemicals, or alternative pest-control agents, can be effectively used. This approach is called "Integrated Pest Management" (see "Risks to Beneficial Insects and Nontarget Animals" in this chapter).

HOW PESTICIDES WORK

If pesticides are chemicals used to kill troublesome living organisms, what prevents them from killing us, too? The answer is: in some cases, very little! One very important factor is that the pesticide should be applied directly to the targeted organism. It is important to avoid incidental contact with the product. The rule is: keep up-wind of sprayers, and wear all the necessary protective gear when applying the chemical.

A second reason people are not nearly so frequently harmed or killed as the target pests is the "route of entry" into the body. This works in a number of ways: pests are usually much smaller than people, so that the area of their surface is greater "per unit weight" (e.g., per gram or per ounce). Small pests often breathe faster, so that they get a far greater dose in a given time. Similarly, the pests (animals, fish, or plants) may have some highly permeable covering or part (e.g., the gills) which allows much easier access for the chemical to enter the body.

Humans normally take precautions to prevent exposure to pesticides in some way. Skin is a reasonably good barrier against water and against many water-soluble chemicals, but many pesticides are more soluble in fat than in water, and so penetrate the skin more easily. Therefore, by using directional spraying; avoiding deliberate exposure to spray or drift; and protecting the skin, we will kill pests, and not ourselves.

With newer pesticides, there is a certain degree of chemical or biochemical targeting. Some early herbicides (e.g., arsenic-based) were very toxic to a great variety of plants, fungi, insects, worms, fishes, birds, and mammals, including people. That is because arsenic attacks a wide variety of proteins, a major part of all living things. This is not a good basis for selective toxicity, because protein is abundant in, and critically important for, all living things. On the other hand, some newer herbicides are aimed at disabling the photosynthetic system of plants. It is recognized, since humans are not green, and do not put on weight just by standing in the sun, that there ought to be much more selectivity in this toxic effect, and those organisms which are not green, due to chlorophyll, should not be affected at all by such a herbicide. This is a very reasonable assumption, and largely true. Nevertheless, there are many steps in the process of harnessing of light to make food, and some of the steps have nothing directly to do with green-colored pigment. So, if the herbicide interferes with photosynthesis at "the wrong step," so to speak, it is entirely possible that some quite distantly related life form may also be damaged.

Nature is very conserving: once she has "dreamed up" some chemical system that is useful as a step in the working of one organism or one life process, she usually hangs onto it, and may use it in organisms as distant as bacteria are from humans. Indeed, all life is related, and so "cross-reactions" to pesticides are more common than we would like.

In a few cases, the biochemical or biological targeting is more precise, and it seems that the target system for the pesticide really is limited to the class of organisms to which the pest belongs — and in some cases, it is specific almost to a single species of, say, an insect. Such recently developed pesticides interfere with chemical communication between individuals in the pest population; interfere with the action of hormones restricted to certain small groups of animals; or interfere with nervous system components or neurotransmitters that are essential for the "pest" to function, but are totally absent or unimportant in people and other nontarget groups. Some of these narrowly targeted pesticides are living organisms, or are derived from living sources. Such living, parasitic, agents are already commercially available; for instance, a microscopic parasite, and a special fungus, have been developed as grasshopper-control agents. These are examples of bioinsecticides. Few other types of pesticides can kill some unwanted insect, while having no effect upon the wanted or beneficial insects, such as the ones that pollinate crops, or make honey.

In summary, pesticides harm the pests much more than they harm ourselves, because of targeting. This means making pests the sole target, and avoiding exposure of other desirable organisms to the pesticide. This can be done in a mechanical way by using protective clothing or avoiding spraying nontarget organisms; by adjusting the dose; and by chemical, biochemical, or biological targeting, through the design of the pesticide.

"CHEMICAL WARFARE" IN NATURE

Many species of plants and animals have used chemicals to kill, disable, or confuse their "enemies," since well before the dawn of history, and some would say, before the advent of humans. The antibiotic drugs we now find indispensable were created in nature many thousands or millions of years ago. Nature did so, not to cure human infections, but to give a competitive advantage to the molds producing them, by killing off the bacteria competing with them for the limited food supply.

The beech tree and the black walnut secrete a chemical called juglone, which prevents the growth of competing plants for several meters around the trunk. The bombardier beetle has a small "cannon" from which it shoots a fuming solution of some caustic material. Many caterpillars spray formic acid from special glands, if threatened. There are Colombian frogs and tropical fishes which produce some of the most toxic compounds ever found. Pyrethrum and the chrysanthemum (both pretty flowers to the admirer of nature) produce potent insecticides (as does the tobacco plant). The wild cherry tree produces compounds that release cyanide in an animal's stomach, to discourage nibbling on its bark or twigs and leaves. It is not humans, but rather nature, that produces the most toxic chemicals known, to maximize the biological gain of one species against another. In fact, the human species is only a newcomer in the history of the use of herbicides, insecticides, and other pesticides!

FORMULATING THE ACTIVE INGREDIENT:
THE FINAL PRODUCT IS A MIXTURE OF CHEMICALS

Quite often a pesticidal product is referred to simply by the name of its active ingredient (e.g., "2,4-D" and "trifluralin"). However, the product is not just a small box or can of the pure active ingredient, but rather a mixture of that active ingredient with a few, or even many, other substances, generally called "inert ingredients" or "inerts." These other ingredients are not pesticides, but have a lot to do with the effectiveness of the product. Consequently, while the active ingredient of a marketed pesticidal product is no secret, the inerts in each brand-name product are, typically, "proprietary" (trade secrets).

THE INERT INGREDIENTS: STICKERS, SPREADERS, EMULSIFIERS, AND COMPANY — INERT TO WHAT?

There are more than 400 different chemicals used as inert ingredients in pesticides formulated in North America. Some of these allow the active ingredient to dissolve in water (solubilizers and emulsifiers); others keep the active ingredient from drifting around as a vapor; still others stick the active ingredient to the plant or animal surface, or help it to penetrate from the surface into the living tissues; and some may prevent foaming or corrosion in spray equipment. Altogether, in the formulated products, these ingredients increase efficacy (i.e., the product's effectiveness in killing the pest). Yet in at least one case, the inerts were more toxic to nontarget organisms than was the active ingredient. Of course, different brands of pesticidal products based upon the same active ingredient can have quite different sets of inert ingredients. On the other hand, totally unrelated active ingredients may be formulated with the same chemical chosen from the many inerts available. The inerts can also differ from time to time in products with identical brand and trade names; this is due to the use of industrially available mixtures which are inherently variable due to the way they are produced, as well as to occasional "improvements" to the inerts.

Exposure to pesticides, then, is not simply exposure to their active ingredients. It is possible that some of the difficulties encountered in trying to better test apparent relationships between pesticide exposure and some illnesses could be due to some of the inert ingredients.

HERBICIDES

Herbicides are chemicals employed to kill plants. There are times when it is desirable to rid an area of every kind of visible plant — for example, in a cracked sidewalk, a parking lot, in some weed-infested industrial area — and then there is little need for specificity. Usually, though, just certain plants ("the weeds") should be killed, while even promoting the growth of others ("the crop").

Some specificity can be introduced by spraying only the weeds (if these are in clumps, or if they are much taller than the wanted plant, for example). In some circumstances, careful timing of the spraying may result in selective kill, even though the agent is just as poisonous to the "crop" as it is to the weed. This is a demanding and rather dangerous solution, because it takes very careful planning, and often requires a large number of workers.

In general, there is today a steady movement toward more selective herbicides. The first significant development along this line was the discovery of 2,4-D and related compounds, which selectively kill broad-leaved plants. This meant that grains and other grasses could be sprayed with such herbicides. As a result, the broad-leaved weeds would be killed, without damage to the (narrow-leaved) crop. Since that time, molecular design of herbicides has sought out chemicals which disable biochemical processes that are found only in plants (e.g., the process of photosynthesis). It is probably fair to say, however, that progression toward highly specific herbicides is still in its very early stages.

Today, there is a large number of herbicides in use in the U.S. and Canada. They can be classified either according to their chemistry, or according to the sorts of weeds they control and the crops they do not damage, or by the way they are applied. In terms of mode of application, some are "preemergent," meaning that they are applied to the soil before the plants emerge from the soil; others are postemergent, where the herbicide is applied after the plants are partly grown. The herbicides may be further divided by method of application into those which are sprayed and those which are used in the form of pellets (which may roll off wanted plants) and, falling to the ground, release their active ingredient, poisoning susceptible weeds.

Herbicides can also be classified by their selective action on various plants. Selectivity of herbicides is a crucial factor in their use in agriculture. There are two types of selectivity recognized in farm and garden use: one is placement selectivity, where the user carefully directs and times the treatment, so that there is very little contact between the herbicide and the wanted crop. The other type is true selectivity, in which both the crop and the weeds can be exposed equally, but only the weeds are killed. This true selectivity is not magic, and a large overdose of the herbicide will kill or severely damage the crop, as well.

Finally, the herbicides can be classified by their chemistry. It is possible to divide the herbicides into just three basic groups, by their chemistry and their source: the naturally occurring herbicides, the inorganic herbicides, and the synthetic organic herbicides. However, these terms are somewhat confusing. For example, the latest synthetic herbicides are sometimes based on modification of natural ones, and nearly all the products in wide use are in the one category: the synthetic organic herbicides. There is a further complication: to the environmentalist or health-food store operator, "organic" means "made without added chemicals." But to the chemist, "organic" simply means "containing carbon." The chemist's organic compounds are more frequently synthetic than naturally occurring.

On the following pages, some aspects of herbicides will be discussed in more detail.

FALSE HORMONES FOR PLANTS

The herbicides 2,4-D and 2,4,5-T share the chemical structure of chlorinated phenoxyacetic acids. The phenoxyacetic structure is similar to one of the true hormones of plants, indoleacetic acid, and the phenoxy herbicides act as false hormones by interfering with the actions of the real hormone.

Under names like "Agent Orange," such chlorophenoxy compounds and some other synthetic herbicides were used as defoliants in the Vietnam War to make targets, hidden under trees, more visible. The earlier synthetic methods of manufacturing these defoliants were later found to generate side-products (contaminants) called "dioxins." It is very important to know that while it is true that one particular member of this family of side-products is regarded as a "supertoxin" because it can kill experimental animals even in minute doses, it is not so clear just how toxic this compound is for humans. In addition, all the other members of the "dioxin" family are less toxic — often dramatically so.

Unfortunately, the term "dioxins," or sometimes even the term "dioxin" is used by the media for such specific names as "2,3,7,8-tetrachlorodibenzodioxin" (TCDD) which is, indeed, a very toxic compound. The important point to remember is that the popular accounts about the toxicity of "dioxins" are chemically so imprecise as to be nearly meaningless. This is not to deny the seriousness of exposure to certain members of that family of chemicals, but to point out that such headlines as "deadly dioxin leak discovered" are impossible to evaluate without much more detail about the exact composition involved.

In the case of 2,4,5-T, the possible contamination with TCDD is the main reason that many jurisdictions have banned its use. More modern methods of production can markedly reduce the likelihood of such extremely toxic side-products.

Even in the absence of any contaminant, large doses of 2,4-D do cause digestive tract disorders, vomiting, diarrhea, and also convulsions or even coma. Recent reviews of the chronic toxicity of 2,4-D are focusing on the possibility that it may produce cancer. For about half a century, 2,4-D has been used in large quantities in many farming areas, and it is only quite recently that research has suggested a possible link to human cancers. The most frequently cited possible connections are to "soft tissue sarcomas," non-Hodgkin's lymphoma, and brain cancer. However, most of the evidence indicates that it is not 2,4-D that causes these cancers.

It follows from this, that abandoning a well-established pesticide in use for several decades, in favor of much newer ones, could easily be to jump from one frying pan into another. This is because any newer substitute simply cannot have a long track record; quite likely not long enough for any connection to cancers or other chronic effects to appear.

TOXIC INTERACTIONS WITH OTHER AGENTS

Those who are prescribed medications are frequently advised to avoid certain foods or beverages, in order to avoid harmful effects of interactions between the drug and other substances (e.g., alcohol). Exposure to some pesticides, as is the case with some drugs, can give rise to similar harmful interactions. Examples may be found in the dithiocarbamate family of compounds, some of which are herbicides (e.g., diallate, triallate), some insecticides and some, fungicides. They are relatively nontoxic to humans, but can produce excessive sensitivity to sunlight in susceptible individuals (with severe sunburns resulting). In addition, they have the ability to increase the apparent toxicity of alcohol, by interfering with its normal pathway of breakdown in the body. In the presence of these dithiocarbamates, alcohol is oxidized only one step, and the product (acetaldehyde) accumulates, causing a poisoning that is characterized by red face, vomiting, and headache. Those people using or otherwise coming into contact with this class of herbicide, should refrain from all alcoholic beverages for at least 24 hours prior to or following exposure.

PARAQUAT: A SPECIAL DANGER

Paraquat is essentially a highly selected dye molecule, having the color of red wine. This resemblance has led to human fatalities when paraquat was transferred from the original container to empty wine bottles for storage. The transfer of any pesticide from the original container to any other must be totally avoided. This applies particularly to containers not conspicuously relabeled or ones that look like those commonly used to store food or drinks. In the case of paraquat, one "swig," even if immediately spat out, is enough to kill a person. The symptoms of paraquat poisoning are ugly: chemical burns of the mouth, esophagus, and stomach occur almost immediately. After about 10 days, the lungs develop fibrous tissue, gradually leading to death by asphyxiation (lack of ability to breathe).

THE INDUCERS: FRIEND OR FOE?

One important effect of even small doses of this class of herbicides is their ability to induce increases of certain enzymes in the liver cells. These enzymes then attack not only this class of herbicides, but also many other compounds foreign to the body, and even some that are normal components of the body, as well. Most of these enzyme increases mean more rapid destruction of the biological activity (including the toxicity) of the synthetic chemicals. Nevertheless, some of the useful and even essential compounds in the body can, at times, be depleted by such overactive enzymes. In addition, some of the conversion products from these enzymes acting upon pesticides or other foreign compounds are, in fact, more toxic than the original compounds. The induction of this set of enzymes consequently explains some of the interactions between two or more pesticides or environmental chemicals in the body, and can both protect us, or place us in greater jeopardy, from future onslaughts of chemicals.

One important group of inducers is the substituted ureas. The term "substituted" as used here is purely a technical term in chemistry. The sort of herbicide referred to includes diuron and fenuron. Such chemicals can produce, with large doses, loss of appetite, reduced activity of the nervous system, lack of coordination, and exhaustion. The substituted urea herbicides are certainly not unique in having this effect upon the "adaptive microsomal enzyme" system, but are just one example of such inducers.

INSECTICIDES

Insecticides are chemicals used to kill insects, or to make them unable to reproduce or develop normally, or, in other words, to make them unable to function as a "pest."

ORGANOPHOSPHORUS COMPOUNDS, CARBAMATES, AND PYRETHROIDS

There are three main classes of chemicals used as insecticides: the *organophosphorus* group (also incorrectly called the "organophosphates"), e.g., dimethoate, diazinon, malathion; the *carbamates*, e.g., carbofuran, carbaryl; and the *pyrethroids*, e.g., cypermethrin, deltamethrin. These three classes of insecticides kill by disabling the nervous system of the insect.

Organophosphorus compounds and carbamates affect the nervous system by preventing the action of an enzyme that normally destroys a specific chemical messenger substance once its message has been passed to the next cell. But in the presence of these insecticides, the messenger molecules persist, build up, and profoundly disrupt the nervous system by continuing to stimulate the receptor cells over and over, instead of just once. The enzyme is called acetylcholine esterase, and it normally removes any "spent" messenger known as acetylcholine.

Since most animals have portions of their nervous systems (nerves and brain) that, like the insects, also operate using acetylcholine, these animals are easily disabled by the same chemicals that kill insects in this way. In fact, humans have the acetylcholine system in many parts of the nervous system too, and if insects were able to spray us with these same "insecticidal" compounds, it would soon be clear that these compounds are quite effective as "peoplicides!"

The third major class of insecticides, the pyrethroids, acts primarily on the nervous system, but not on the "messenger" or neurotransmitter systems. The pyrethroids are thought to act upon the membrane structures characteristic of the nervous tissue.

These three major classes of insecticides illustrate two features of toxicology. One is that the relative seriousness of the biological impact of a chemical used often depends upon the particular organism (plant, animal, microbe, or humans) we are interested in. If we ask the question, "Which is the most toxic of the three classes of major insecticides?," it seems perfectly reasonable to expect a simple answer. In fact, however, the organophosphorus group is by far the most toxic to humans (followed by the carbamates), but the least persistent in the environment. On the other hand, the pyrethroids are not very toxic to humans, but extremely toxic to aquatic invertebrates (lower aquatic animals) and some are persistent. So, there is no simple answer to the seemingly straightforward question.

The second feature of toxicology that is illustrated by some of the common insecticides is the ability of the very same compound to produce different or related sets of symptoms, in the same species (and even in the same individual) by different ways and means. Different series of steps may lead from the same toxic chemical to two different outcomes. It is like dividing a stream right at its source. In the case of the organophosphorus insecticides, many produce the immediate disruption of the nervous system we have just considered —

and only that effect (e.g., parathion, malathion). Some other members of this class of chemicals produce, in addition, a delayed disturbance of only the nerves, not of the brain (e.g., leptophos, mipafox). There are other organophosphorus compounds (e.g., triorthocresyl phosphate) which cause just delayed disease in nerves, but little or no immediate effect upon the "messenger" function of the nervous system.

PHEROMONES AND HORMONE MIMICS

Besides the main groups of insecticides mentioned before, there is a fourth group, which is really more of a collection, with very different kinds of chemical structures, rather than one group. These include some pheromones (compounds which act like hormones, but from individual to individual, rather than within the individual). Others mimic the true or internal hormones of insects. The mechanisms by which these chemicals act, and the chemical structures themselves, are not shared by us or other mammals, and hence, there is a much better specificity or chemical targeting of these insecticides.

The first minor group, the pheromones, are rather specific, and can be used to lure particular kinds of insects (pests) to traps, while having no effect upon such useful insects as the honeybee.

One example of the use of chemicals related to insect hormones (the second minor group) is the application of molecularly altered "juvenile hormone" to mosquito larvae, preventing them from becoming adult, flying, humming, biting "pests."

BIOLOGICAL CONTROL AGENTS

These control agents are biological, rather than chemical. Their greater specificity is due to the biological means of targeting, which may include parasitic action or specific means of entering certain pests. Some of these biopesticides, in the final analysis, act by poisoning the target species with toxins. An example already on the market is the bacterium *Bacillus thuringiensis* (*B.t.*). Other living organisms are now being tested or developed for use in controlling grasshoppers.

WHAT ARE THE CHARACTERISTICS OF AN IDEAL PESTICIDE?

The very best of pesticides would:

- Prevent the problem posed by the pest (by killing the pest, if necessary, or by preventing its reproduction, development, or impact).
- Be specific — that is, be effective against one or more pest species, but without harmful effects upon nontarget organisms (crops, people, pets, wildlife, fishes, and other aquatic life, wild plants, bees, beneficial soil and water microorganisms, etc.).

- Persist just long enough to be effective in controlling the specific pest(s), but not long enough to build up from application to application, or year to year; and not long enough to move into areas away from its original application.
- Be readily degraded or metabolized (changed chemically, through the action of living organisms or the environment), in such ways that the products are nontoxic.
- Be free of any harmful interactions with other substances frequently found in foods, feeds, or part of the human lifestyle.
- Be easy to apply, without risk of drift or volatilization (boiling-off on hot days).
- Have a known mechanism of action on living things, and a known antidote (method of treatment).
- Produce no interference with the freedom to rotate crops in the same treated area.
- Be readily marketed in a fully recyclable, or biodegradable, or dissolvable package.

SOME ENVIRONMENTAL ASPECTS OF PESTICIDES

Earlier in this chapter, the nature of pesticides and both their benefits for and toxicity to people were presented. In this section, the focus is shifted to the environmental toxicity of pesticides and to ensuring the safe use of these agents. Their safe and responsible use, encouraged by legislation, but ultimately accomplished through the actions of the agricultural chemical industry and the individual farmer, ensures the protection of the ecosystem and safety of our food supply.

RISKS TO BENEFICIAL INSECTS AND NONTARGET ANIMALS

Many insects are pests, destroying crops, transmitting diseases, or distressing people and domesticated animals. Yet many more are beneficial, pollinating plants, helping to control pests by eating them, or producing food for people (principally honey). Since nearly all synthetic (and some natural) insecticides kill insects by attacking their nervous system enzymes or membranes, there is practically no selectivity. The selectivity must be practiced by the user, through a combination of control of the areas treated and the time of day when treatment is applied. Not all insects are equally vulnerable to spray at a given hour of the day; they may forage over different periods. Much greater selectivity can be obtained by using bioinsecticides (e.g., parasitic single-celled organisms or fungi, specific predatory animals, bacterial strains causing disease in only a few insect species) and combinations of such biological agents with other means of insect control, in what is called Integrated Pest Management.

The risks from membrane-affecting insecticides are not as widely spread as are the risks from the enzyme-targeting insecticides. The membrane targets are shared mainly with the small aquatic invertebrates, but not fishes, birds or mammals; on the other hand, the enzyme targets of the other major class of insecticides are shared, to some extent, by practically every type of animal, including humans.

ECOTOXIC EFFECTS: DIRECT AND INDIRECT

Toxic effects upon the ecosystem and among wildlife are of two different kinds. Direct effects are those in which the toxic agent is absorbed into the affected organism and has a harmful effect in that organism. Indirect effects are those in which some agent has a harmful effect on one component of the ecosystem. The initial effect (the death or disappearance or diseased state of that organism) reduces the food or nesting sites or other aspects of the habitat required for a second species, which either does not receive, or is tolerant or resistant to, the chemical applied.

PROBLEMS WITH CHANGING USES OF HERBICIDES

Some pesticides (e.g., the herbicide glyphosate), like some growth regulatory substances, now can bring a crop quickly to maturity and allow efficient harvest with reduced losses. Broad-spectrum herbicides for which there are corresponding herbicide-resistant crop varieties, can be used to totally eradicate all noncrop plants from any sprayed area. Substituting herbicide spray for conventional tillage (sometimes referred to as "chem fallow") conserves both moisture and energy. The flip side of these advantageous practices, however, is the high probability that large amounts of such agents will now be applied, resulting in a far more complete monoculture in the fields and accidental destruction of hedgerow, fence-row, field-margin, and roadside wild plants and wildlife habitat. Adopting such newer techniques of farming will need to be paralleled by very careful avoidance of damage to the embedded and adjacent, richly diverse, plant communities and habitats.

THE OUTMODED AND THE OUTLAWED:
ARE THEY REALLY GONE?

It is not uncommon to encounter pesticidal materials, now outmoded, prohibited, or restricted in developed nations, being used in a general and widespread way in some of the developing or so-called "Third World" nations. Surprise is often expressed over the continuing findings of residues of DDT or its breakdown product in North American plants, people, and birds — surprising because DDT has been banned in North America for a couple of decades. On the other hand, there is long-range transport of pesticides in the moving air masses of the world; also many birds migrate each year, through the tropics, to more northern areas. These migrants overwinter in countries where DDT is still very much in use, and where the risks associated with not using it far outweigh, in conventional wisdom, the other risks, namely, to the environment. In fact, the World Bank has recently provided funds for setting up three factories to manufacture DDT within developing countries.

ENSURING PROTECTION FROM ANY HARMFUL EFFECTS OF PESTICIDES

Safe use of pesticides and full protection from their potential harmful effects involve actions by all of us: the government agencies registering and monitoring these products; the agricultural chemical industry developing, making, and marketing them; and the individual actually applying pesticides, or consuming food produced with their use.

ACTIONS BY GOVERNMENT AGENCIES

Government departments and agencies are given the responsibility to ensure the safety of the many pesticides, in terms of human health (occupational, user, bystander, and consumer) and the health of the environment (wildlife, nontarget plants, habitats, and ecosystems). This protection typically involves federal-level departments of health, labor (human resources), environment, and agriculture, in association with transportation, and fisheries (along with their specialized agencies), and often state or provincial and territorial counterparts. The legislative bodies enact laws, draft regulations, and put in place systems for monitoring the situation, researching new developments, and prosecuting offenders. There is room to mention only some of these laws and agencies here; for a more comprehensive coverage of the governments' roles in ensuring pesticide safety, see the book by Kamrin listed in the "Further Suggested Reading" section of this book.

For those who are interested in the policing of pesticides in food, more information can be obtained from studying the Federal Food, Drug and Cosmetic Act in the U.S. and the Food and Drug Regulations in Canada.

In the U.S., pesticides in general come under the Federal Insecticide, Fungicide and Rodenticide Act (FIFRA), but the most pertinent legislation relating to the topic of pesticide residues in foods is found in sections 408 and 409 of the Federal Food, Drug and Cosmetic Act, including its Food Additives Amendment and Pesticide Chemicals Amendment. The Environmental Protection Agency, mandated by Section 3 of FIFRA, sets the U.S. limits ("allowable levels") for pesticide residues, but these are enforced by the Food and Drug Administration (FDA). Actual levels of pesticide and other chemical residues in foods are checked through the U.S. National Monitoring Program for Food and Feed. This program has three components: the Total Diet Study of market foods (by FDA); nationwide monitoring of unprocessed food and feed (by FDA); and analysis of meat and poultry products (by the U.S. Department of Agriculture).

In Canada, the Food and Drug Regulations specify the maximum residue limits in food for quite a number of agricultural pesticides. The limits imposed include, in most cases, at least a 100-fold safety margin. For pesticides not listed in the Canadian regulations, for example, the allowable residue limit is automatically 0.1 ppm. In the case of Canadian food, the continuous monitoring carried out by Health Protection Branch within Health Canada, Ottawa, indicates that most samples (85%) do not show any detectable residues. The comparable food monitoring in the U.S., carried out under the National Monitoring Program for Food and Feed, has had similar findings.

ACTIONS BY THE AGRICULTURAL CHEMICAL INDUSTRY

It rests with the agrochemical industry to develop active ingredients and to formulate mixtures which are effective, optimally specific, and as nontoxic as possible to both people and the rest of the ecosystem. This industry is required to provide experimental proof of minimal toxicity to nontarget species, submitting this to a set of government agencies for review, whether for new products or for products already registered, but scheduled for periodic review. In both the U.S. and Canada, associations of producers of agrochemicals have adopted progressive policies that go beyond those strictly required by government. These industry initiatives include "Product Stewardship" programs (in which a cradle-to-grave management of product is assured) and "Responsible Care" programs (in which a comprehensive set of corporate operating principles is made public, and to which companies must subscribe and conform, in order to display the logo of this program). The agrochemical industry in both the U.S. and Canada has recently completed large-scale research into the reengineering of their pesticide containers, to build in features for safer dispensing of the contents, and to ensure recyclability. At present a consortium of agricultural chemical companies, paint companies, and other chemical producers has a massive study under way to more fully test the adequacy of protective gloves, now that less-common solvents are being proposed as formulating agents in commercial products.

ACTIONS BY THE INDIVIDUAL

Individuals using pesticides fall into three groups: commercial applicators, farmers, and the general public. While the amount of pesticide applied decreases in the above order, the amount of training in safe use of pesticides, and the precautions taken, also decrease in the same order. Most jurisdictions require that commercial applicators be trained, certified, and licensed in the safe and appropriate application of pesticides. Several jurisdictions also require that farmers be similarly trained and certified, in order to purchase the large quantities of pesticides used on many farms. On the other hand, about 11% of all pesticides are used "in the back yard," without the benefit of any training in their safe and appropriate use.

Beyond their own proper, minimized, and targeted use of pesticides according to the label directions, the general public should be aware of pesticide use in their vicinity; knowledgeable about alternatives; and careful both in their selection and preparation of produce for eating and in laundering clothing exposed to pesticides.

To Wash, Scrub, Pare, Peel, Cook, or Discard?

Produce that is marketed is expected to be used in a conventional manner. This assumes, as far as the safety of foodstuffs is concerned, that pesticide-treated grapefruit, orange, lemon, or even apple skins are not expected to be eaten, but some people use them to make marmalade, candied fruit, or just like to eat the peels.

Warning statements that are required by law or voluntarily placed on boxes of produce often cannot be seen by consumers unless they are buying directly from the original box, where the whole box is exposed to view, with the warning visible. It is scarcely expected that the warning will be attached to each apple, orange, or kumquat!

Research has shown that first washing produce with warm or hot water, and then peeling and cooking it, together remove most of the residues. Therefore, it is a wise practice to wash, scrub, or peel produce.

Pesticides and Laundry

You might think that after you have taken all the precautions, applied the pesticide according to the label, have worn the proper protective gear, and cleaned up after the operation, that there is nothing else that remains to be done in order to minimize toxic hazards from the day's work. But if you applied a lot of pesticide (e.g., on a farm, or as a commercial applicator) there is still one critical step to be taken to ensure your health and that of your family. This is the laundering of the work clothes. Fortunately, some research has been done recently into the effective ways of removing pesticides from clothing; unfortunately, these same studies have shown that the average family uses few of these. To ensure the removal of all or nearly all the pesticide, the following steps should be followed:

- Change work clothes daily.
- Keep the clothes, used during pesticide application, separate from the usual family laundry. Use plastic bags to ensure this separation.
- Use a "pre-spray" laundry aid (Spray 'N Wash®*, Stain-Away®†, etc.) before starting the wash with detergent.
- Make sure the water heater is set to the hottest temperature that is safe, and not turned down to conserve energy. Many families use wash water that is not hot enough for removing the pesticides.
- Use the full amount of detergent recommended on the box. Some families cut down on the quantity of detergent for their ordinary laundry, but such a "saving" will not work for clothes that must be freed of pesticides.

---

*   Registered trademark of DowBrands L.P. (Household Products Division). In U.S.: Indianapolis, IN.

†   Registered trademark of Colgate-Palmolive Canada Inc. In Canada: Toronto, ON.

- Wash the clothes three times before reusing them, if the use of pesticides was substantial (e.g., commercial applicators and farmers).
- Dry clothing outside, not in the dryer.
- When selecting clothing to wear while applying pesticides, it is best to avoid "perma-press" type fabrics, because these are harder to launder to a pesticide-free condition.

For more than occasional users of pesticides, proper laundering is an important part of dealing with these chemicals safely. It makes little sense to put on yesterday's pesticides when starting a new day "in the fields."

Chapter 5

OTHER CHEMICALS INVOLVED IN PRODUCING FOOD

Having discussed the pesticides used in the production of food, it is worthwhile considering other products that can also be involved in food and livestock production.

FERTILIZERS

Farm fertilizers are not commonly thought of as toxic or hazardous, but some of them are. Both nitrate fertilizers and anhydrous ammonia pose considerable explosive hazards: the former, from possible ignition; the second, because it is stored and transferred under very high pressure. In terms of toxicity, anhydrous ammonia is corrosive to eyes, mucous membranes, and skin, and it damages the lungs if inhaled.

Fertilizers containing nitrate can be dangerous to cattle, because nitrates can be converted to nitrites in the body. These nitrites, then, disable the oxygen-carrying capacity of the hemoglobin of the blood, causing rapid death. Such poisonings occur when cattle have a chance to eat from a fertilizer bag, but more often nitrate poisoning is associated with the drinking of nitrate-containing water, or occurs after eating large quantities of fodder which happens to contain lots of nitrogen and nitrate.

Nitrogen fertilizers also contribute to other problems, such as eutrophication (overenrichment) of water bodies, and indirectly are a threat to the ozone layer in the upper atmosphere, which shields all living things from too much radiation from space. Nitrous oxide, produced by bacteria from nitrates and nitrites in soil and water, is oxidized in the upper atmosphere to ozone-destroying nitric oxide. The more nitrogen fertilizers are used, the more nitrous oxide will be released, and this means that more nitric oxide will be in the atmosphere, and less ozone. Therefore, it may be worthwhile to consider using more natural fertilizers (manure and compost), because they are said to not "leak" as much nitrogen into the atmosphere. Of course, nitrogen fertilizers are not the only source contributing to the destruction of ozone. Nitrogen oxides are emitted in large quantities by coal-fired power plants, and other chemicals such as chloroflurocarbon-type propellants formerly used in spray cans, can also destroy ozone.

Phosphate fertilizers are not particularly toxic, except in that they can contribute to eutrophication of water bodies and can contain "stow-away" or "piggy-backed" trace metals such as cadmium. Such ecotoxic concerns will be covered in Chapter 10, "Industrial Chemicals," and Chapter 11, "Waste Chemicals."

GROWTH REGULATORS

When mass-producing crops, it is often beneficial to use an agent which will slow down, speed up, or just synchronize ripening. Such agents give more control over harvest times, may help avoid frost damage in areas with short growing seasons, and allow harvesting in one pass, reducing fuel consumption and land compaction. Chemicals used specifically for these effects are not pesticides or fertilizers, but belong to the general class called growth regulators. The most publicized growth regulator is Alar®*, used, until recently, to keep apples on the tree longer, so they would develop a better color and greater crispness before harvesting; it also made the apples ready for a synchronized and efficient harvest. In 1989, the Natural Resources Defense Council (NRDC) suggested, in a widely publicized commissioned report, that Alar increased cancer risk, particularly among children, who eat a lot of apples and applesauce. This publicity alarmed parents, resulted in a massive drop in sales of apples, government de-registration of Alar, and withdrawal of the product by its manufacturer. However, very few toxicologists were convinced of the validity of such claims, and the actions taken were driven far more by fear than by scientific evidence. It is not easy to fully evaluate the NRDC report because the toxicity of the active ingredient in Alar changes if the apples are cooked, and because the basic data are for rodents, not people. The "Alar scare" probably did some good in that it made government regulators everywhere look again at the differences between childhood vs. adult susceptibility and food habits, in relation to chemical residues in foods. It certainly produced many financial losses in the apple industry, and a former executive of EPA at the time is reported to since have called the NRDC report "gravely misleading," adding that the public was prone to give credence to that report, to believe the worst, despite EPA's own counter-statement at the time. A highly critical assessment of the Alar story may be found in the book by Ray and Guzzo (1990), entitled *Trashing the Planet*. As in any such polarized issue, perhaps the most accurate view lies somewhere closer to the middle ground.

---

* Registered trademark of Uniroyal Chemical Ltd. In Canada: Elmira, ON.

FEED ADDITIVES

The success of rearing large numbers of animals in confined spaces and at the same time producing good weight gains, is due in part to the use of feed additives, including growth promoters and antibiotics. Intensive livestock production is probably here to stay, due to consumer demand for affordable, wholesome food. However, some public concern has arisen regarding feed additives and growth promoters. It is appropriate, then, that they be considered here.

Before doing so, a fundamental distinction must be made: use of veterinary drugs to treat disease as opposed to use in growth promotion. Some diseases in animals must be treated with drugs. Veterinarians are entitled to prescribe drugs, be it for injection into the animal directly or for application in the feed. However, such prescriptions are like prescriptions written by a physician — they have a limited time span and carry clear instructions such as: "apply x times daily for y days." In all these cases, it is also necessary to specify at what date (after administration of the medication) the animal products may be fit for human consumption. Such prescriptions are regulated: in the U.S. under the Federal Food, Drug and Cosmetic Act (and its Animal Drug Amendment), administered by the Food and Drug Administration, and in Canada under the Food and Drugs Act, administered by Health Canada.

A totally different matter is the use of drugs in feeds, for growth-promoting purposes. In the U.S., antibiotics and certain other animal drugs are used in treatment of diseases, for growth promotion, and in prophylaxis or prevention of possible illness among the food animals. In the U.S., the various aspects of veterinary drugs and feed additives for food animal production are governed by a number of acts and the regulations and activities they mandate. The Animal Drug Amendment of the Federal Food, Drug and Cosmetic Act requires premarket clearance of all new animal drugs and feeds containing them, through the Center for Veterinary Medicine of the Food and Drug Administration. The Food and Drug Administration routinely monitors animal feed, while the U.S. Department of Agriculture analyzes meat and poultry products for residues of various chemicals.

In Canada, the Compendium of Medicating Ingredients Brochures lists those medicating ingredients permitted by Canadian regulations to be added to livestock feed. To comply with the Feeds Regulations, all medicated feed manufactured, used, or sold in Canada must be prepared in such a way as to adhere to the specifications listed. The use for disease treatment (therapeutic use) is included. These regulations are administered by Agriculture and Agri-Food Canada. Included are such things as species, purpose, dosages, and, when necessary, withdrawal times, i.e., time span from last use to slaughter of the animal. According to the Canadian regulations, drugs can be offered and sold for growth promotion, but the levels of drugs allowed to be used are much lower than those used for treatment of diseases. Withdrawal times for use in growth promotion are either not needed or are very short compared to withdrawal times for therapeutic drugs.

While there is little question that growth promoters are effective in boosting animal productivity (therefore increasing economic benefits to the producer), there is considerable discussion as to whether such practices are really sound and safe. Some people question whether it is absolutely certain that no growth promoter ends up on our dinner plates. This is an issue that is hotly debated today. However, to be fair to producers and consumers, the following has to be said:

- There is little, if any, direct evidence that any feed additive or growth promoter (properly used) has caused any direct health effects in people.
- The Institute of Medicine of the U.S. National Academy of Sciences established a special committee to conduct a "Human Health Risk Assessment of Using Subtherapeutic Antibiotics in Animal Feed." Their report states that "The committee believes that there is indirect evidence implicating subtherapeutic use of antimicrobials in producing resistance in infectious bacteria that causes a potential human health hazard...." The committee estimated that about one third of the subtherapeutic antibiotic use in animals was for growth promotion — about one million kilograms per year, of those general classes of antibiotics also used in treating human diseases. However, it is important to recognize that it is only a minority of individual antibiotics within these classes which are actually used both in growth promotion and in treatment of human diseases.
- Human error (preparing or using the wrong concentrations of a feed additive) is likely to occur on occasion.

One possible solution would be to offer for sale some meat carrying the label: "Guaranteed free of growth promoters." However, such products would require a higher price tag. Perhaps it would be more productive to recognize the generally very high safety of livestock products in the U.S. and Canada; to remain vigilant, supporting the difficult and highly technical research (plasmid-tracing) necessary for further progress in this field; and to encourage producers to work still more closely with their veterinarians to prevent problems from antibiotic residues or bacterial resistance.

TOXIC SUBSTANCES UNINTENTIONALLY RESULTING FROM FOOD PRODUCTION

In addition to pesticides, fertilizers, growth regulators, and feed additives used in modern farming, there are a few other chemicals on the farm that are toxic but not placed there deliberately. The main ones are gases arising from biological processes occurring in manure or in silage. Every year significant numbers of farmers die from being overcome by these gases. In addition, while not toxic in the most common sense, ammonia emanating from feedlots and excess fertilizer run-off from farms both can cause problems in ponds and nearby habitats, a process called eutrophication. One of the effects of this over-abundance of nutrients is excessive growth of some types of microscopic plants called algae. When the algae die, the process of their decay strips oxygen from the water to such an extent that fish or other important aquatic life cannot breathe and then die. Some algae also produce toxins which have been known to kill dogs or cattle or cause disease in humans, when these toxins are taken in while drinking water.

While farming operations of various kinds can produce considerable amounts of ammonia and smoke, with their ecotoxic effects and possible long-term human health effects, the main toxicants produced on some farms (from the point of view of immediate effects on people) are the two gases hydrogen sulfide and nitrogen dioxide.

Hydrogen sulfide is formed from manure by bacteria. The gas smells strongly like rotten eggs. So, one might expect no problem — when the smell gets bad, surely the farmer or his family would get out of the affected area. This is not the case. Hydrogen sulfide does "stink," at low concentrations, that is true. But, at a higher concentration, the gas numbs the sense of smell, and thereby cannot be noticed. Hydrogen sulfide is far more toxic than is generally believed — it is almost equal to cyanide. It is also quick acting, and not easily reversed, once a large dose has been inhaled. Certainly the first step in treating someone overcome by this gas is to remove that person from further exposure to it. But, this can be done safely only by someone wearing a self-contained breathing apparatus. Far too often, a son has made a heroic effort to rescue his father, found to be overcome by hydrogen sulfide, only to produce a double tragedy, with the son dying as well from this toxic gas.

When silage ferments, carbon dioxide and nitric oxide are formed; the latter is then oxidized to nitrogen dioxide. The mixed gas has been shown to be responsible for a condition in humans known as "silo filler's disease." Although the initial signs may be only a mild irritation, death may occur up to a month later from lung injury. In other cases, repeated inhalation of the gas can lead to permanent changes in the lungs, known as emphysema. Animals (cattle, pigs) can be similarly affected.

The ecotoxic effects of excess ammonia from some intensive livestock operations (pig barns, feedlots), both directly and through conversion to nitrates and nitrites, are covered in Chapter 11, "Waste Chemicals." The various toxic effects of smoke are covered in Chapter 13.

Chapter 6

FOOD, FOOD ADDITIVES, VITAMINS, AND MINERALS

FOOD

Many years ago most people ate home-cooked foods and locally grown fruits and vegetables in season. There were few, if any, fresh fruits and vegetables available all year long. Today, in developed countries, fresh produce, meat, seafood, and an extensive array of canned, frozen, processed, and prepared foods from many regions are readily available throughout the year.

In developed countries, life expectancy has risen, so that most people now live beyond the age of 70. In the early 1900s, people died from infectious diseases such as tuberculosis or diarrhea. Today, heart disease, cancer, or cerebral vascular diseases (e.g., stroke) are more likely to be the cause of death. In fact, most physicians and nutritionists are more concerned about our over-consumption of fats, salt, and alcohol than they are about malnutrition.

Most developed countries have safe food supplies, yet people are confused about, and fearful of, the wide range of chemicals in food. Some are also concerned about food additives, pesticide residues in produce, drug residues in meat, growth regulators (e.g., Alar® in apples), contaminants, and improper packaging. Worry is widespread that synthetic pesticide residues in the food supply may cause diseases such as cancer. Our distress has been accentuated by the continuing stream of new research findings. Experts have disagreed whether artificial sweeteners cause cancer or not and even dispute the question whether excessive intake of cholesterol contributes to cardiovascular disease. It has been suggested that a diet high in saturated fats (which are used by the body to produce cholesterol) is of greater concern than cholesterol alone.

There is less concern about naturally occurring chemicals, yet thousands of natural pesticides have been discovered, and every species of plant contains its own set of chemicals. It is evident that nature is not benign; it has been estimated that about 10,000 times more of the naturally occurring pesticides are eaten than synthetic ones.

The human diet contains a great variety of naturally occurring mutagens, antimutagens, carcinogens, and anticarcinogens. For example, foods such as cabbage, bananas, celery, and carrots (the list is too long to name them all) contain all sorts of naturally occurring chemicals with biological activities and strange sounding names. Yet these foods are good for us.

Despite the array of naturally occurring and synthetic chemicals which might be present, the food supply can be considered safe. In general, our bodies are able to handle both the naturally occurring and the synthetic chemicals, in their usual trace (or lesser) amounts.

FOOD ADDITIVES

Food additives are substances incorporated during the production, storage, processing, or packaging of foods. They can be of natural or synthetic origin. In addition, they are added to food either intentionally (direct food additives) or unintentionally (indirect food additives).

It is difficult to quarrel with the necessity for additives to preserve processed foods during their manufacture and shipment and while they sit on grocery shelves for a reasonable period of time. Nor can we contest their usefulness if they enhance flavor or appearance, create satisfactory and appetizing textures, and so on, as long as they are safe.

DIRECT FOOD ADDITIVES

Direct food additives are added during processing to perform a specific function, e.g., to preserve or improve the real or perceived quality of the food or to aid in processing. Examples are antioxidants (which prevent rancidity), coloring and flavoring agents, vitamins, minerals, and inhibitors of bacteria and molds. Emulsifiers (which improve processing, texture, and handling) are widely used additives. Flavoring agents, chemically a very diverse group, comprise the largest single class of chemical substances used in food preparation. Among the oldest food additives are salt, spices, and wood smoke.

A number of direct food additives have been suspected to cause toxic effects. Administration of the antioxidants, butylated hydroxyanisole (BHA) and butylated hydroxytoluene (BHT), has produced changes in the livers and kidneys of laboratory animals and affected their reproduction. Ingestion of nitrites (used to cure meat and often found in cheese) may contribute to the incidence of stomach cancer. Sulfites, as preservatives on fresh fruits and vegetables in restaurants, have caused so many allergic reactions in people that sulfites have been banned for such purposes. However, the sulfite compounds can be used legally as preservatives in dried or frozen fruit, fruit beverages, wine, beer, jams, and jellies.

Some food colors have been associated with cancer and reproductive problems (amaranth, i.e., Red Dye #2) and learning and behavioral problems in children (tartrazine).

Monosodium glutamate (MSG) is a flavoring agent associated with an allergic reaction termed "Chinese restaurant syndrome." People sensitive to MSG experience headache, numbness, tingling of the mouth and tongue, and weakness after eating "Chinese" food. Monosodium glutamate may also affect brain development, so it is not added to food intended for children under 1 year of age.

However, the last example, monosodium glutamate, points to a very important fact: Not all people react the same way. Some do not react to certain chemicals at all, others do so on occasion, while another group experiences severe reactions frequently, if not regularly. The people in this latter group consider themselves "allergic."

For such allergic people, and also for many others who simply wish to avoid certain or all additives, it is important to know whether (and exactly which) additives were used in a specific food product. Many packaged items on the grocery shelf carry labels indicating the ingredients, in descending order of amounts. Such a label might read:

> Sugar (may also contain dextrose), corn dextrin, vegetable oil, shortening, citric acid, tricalcium phosphate, trisodium citrate, natural lemon flavor, food color, vitamin C (214 mg per 100 g), calcium oxide.

In most European countries restaurant menus indicate which additives or preservatives have been used. Consumers everywhere should insist on as much information as possible, so that they can make the right choices.

INDIRECT FOOD ADDITIVES

Indirect food additives are unintentionally added to foods during production, processing, storage, or packaging. They can occur in trace amounts and are permitted in foods only if they cannot be avoided by "good agricultural and manufacturing practices."

Three types of indirect food additives of concern are industrial chemicals, pesticides, and drugs. Industrial chemical and pesticide residues are sometimes detected in trace or small quantities in many foods. There is no conclusive evidence of poisoning from consuming such food. Poisonings have occurred from residues when pesticides were applied incorrectly (for example, when fruits or vegetables were harvested too soon after treatment with pesticides), and when fungicide-treated seeds were eaten by people or livestock. Drug residues should not occur. They are avoided when producers follow correct application schedules. Constant surveillance and frequent "food basket" surveys by government agencies, to enforce regulations, continue to be needed to protect the consumer.

VITAMINS AND MINERALS

Much has been written about megavitamin therapy, and the topic remains controversial. However, it can be said that vitamins A, B_6, D, and E, and the minerals iron, selenium, and zinc are potentially toxic.

Prolonged consumption of two or three times the normally recommended dose of vitamin A can adversely affect a number of organs. These include the bones (fragility, stunted growth), central nervous system (headache, loss of appetite, hydrocephalus), gastrointestinal tract (nausea, vomiting, pain), skin (dry lips, rash, hair loss), and the liver and spleen (enlargement).

Excessive doses of vitamin B_6 (pyridoxine) have been advocated for some psychological disorders. However, suddenly stopping these large doses can cause convulsions. Daily ingestion of large doses of vitamin B_6 for several months has resulted in nerve tissue toxicity in humans. Complete recovery after the drug has been stopped may not occur for 2 or more years. Not all clinical studies show these negative effects.

Excessive intake of vitamin D produces vomiting, diarrhea, weakness, and weight loss. A more significant problem with excess vitamin D is excessive calcium absorption. This can lead to calcification of soft tissues such as the kidney, lungs, and blood vessels.

Iron can be a problem in terms of childhood poisoning. A single bottle of iron tablets (used to treat anemia) often contains enough iron to fatally poison a child. Fatalities in children have occurred from as little as 400 mg elemental iron. Medical assistance should be sought immediately if ingestion of medicines containing iron is suspected.

Selenium and zinc supplementation should not be necessary if a person is eating an adequate diet. It is very questionable whether selenium can protect from or "cure" a variety of disorders including cancer, heart disease, arthritis, sexual dysfunction, hair problems, and aging. In fact, excessive exposure may cause brain damage with headaches, dizziness, and convulsions. Excess zinc exposure causes gastrointestinal problems.

The use of the antioxidants beta-carotene, vitamin C, and vitamin E has been advocated to prevent some types of cancer, heart disease, osteoporosis, and cataracts. Whether taking these supplements is beneficial to prevent these diseases is controversial. It might be wise, instead, to supplement one's diet with foods rich in these antioxidants, such as green, leafy vegetables and fruits and vegetables of a deep orange color, because of their beta-carotene. Adequate folic acid, another vitamin B compound, taken by a pregnant woman, has been shown to lower the risk of one type of abnormality (neural tube defects) in the fetus.

In most cases, a well-balanced diet will supply the nutrients a person needs. However, certain segments of the population (e.g., pregnant women, the elderly, or the young, depending on their nutritional status) may require supplementary vitamins and minerals.

Chapter 7

THE MEDICINE CABINET

Although the medicine cabinet may contain a number of prescription and "over-the-counter" medications, only the latter will be discussed here. A number of books are available concerning such drugs, a few of which are listed (see books by Berube, the Canadian Medical Association, and the U.S. Pharmacopeial Convention, Inc.) in the section "Further Suggested Reading."

Over-the-counter medications are not considered harmful to most people when taken at recommended doses. However, for many of these products, a single bottle contains enough drug to poison a child, often fatally. Add to this the hazard of other items in the cabinet, such as antiseptics or astringents (iodine, hydrogen peroxide, and rubbing alcohol), and it is easy to see how the medicine cabinet can be the source of many poisonings. Here, we shall discuss only a few of the products (e.g., cough and cold preparations, analgesics, stimulants, sleep aids, antacids) likely to be found in anyone's medicine cabinet.

COUGH AND COLD PREPARATIONS

Common cough and cold preparations contain one or more of the following: antihistamines, nasal decongestants, and antitussives. Antihistamines help to decrease the amount of mucus secretion and thereby provide some relief from a runny nose. Other uses of antihistamines are to relieve symptoms of hay fever and other types of allergy.

The three major nasal decongestants used in cold and cough preparations are pseudoephedrine, phenylephrine, and phenylpropanolamine. Nasal decongestants are used to relieve stuffy nose (nasal congestion) and other symptoms of colds and hay fever. Antitussives, such as codeine and dextromethorphan, are added to cough medicines to suppress a dry nonproductive cough, but should not be used for productive coughs.

The most frequent side-effects seen with cold and cough preparations are dryness of the mouth, nose, or throat, and drowsiness. On the other hand, medicines containing only a decongestant may stimulate the central nervous system.

Cold and cough preparations should not be taken at the same time as alcohol or other drugs which affect the central nervous system. Their effects are potentiated (increased) by these types of agents. Since some of these medicines cause drowsiness in some people, it is also important to know how you react to them before driving a vehicle, operating machinery, or doing any jobs requiring that you be mentally alert. These medications should also not be taken by people with certain health problems (e.g., high blood pressure). Check with your physician or pharmacist first.

ANALGESICS

The analgesics (painkillers), acetylsalicylic acid (ASA; Aspirin®) and acetaminophen (Tylenol®*), are among the most widely used nonprescription drugs. More recently, ibuprofen (Advil®†; Motrin®**) became available. All these drugs relieve pain and fever, but only ASA and ibuprofen reduce inflammation.

Acetylsalicylic acid and ibuprofen are considered safe when taken at recommended dosages and for short periods of time. However, effects on the gastrointestinal tract (stomach-ache, heartburn, nausea, and bleeding) and effects on the blood (decreased clotting) may occur. Therefore, these medications should be used with caution by people with stomach ulcers, a history of blood coagulation problems (such as hemophilia), diabetes, gout, or asthma. People with asthma may develop extreme allergic reactions.

The advantages of using acetaminophen are that it does not cause gastric bleeding, and can also be used by people who are allergic to ASA and other such drugs. Nevertheless, it can cause severe liver damage in overdose, or kidney damage if taken chronically.

STIMULANTS AND SLEEP AIDS

Stimulants promote alertness and decrease the sense of fatigue. Caffeine is a natural stimulant present in coffee, tea, and some soft drinks. The concentration of caffeine in coffee is 100 to 150 mg/cup and is within the range to stimulate the central nervous system. Toxicity can develop following ingestion of 1000 mg or more of caffeine at one time. Effects include insomnia, restlessness, excitement, twitching muscles, rapid breathing, and increased heartbeat. Caffeine is the chemical most frequently found in "wake-up pills."

One common sleep aid is diphenhydramine. Overdose in adults results in nausea, vomiting, diarrhea, and drowsiness. This is followed by excitement, convulsions, and slow, shallow breathing. Consumption of alcohol or other medications that "depress" (slow down or decrease) the functions of the central nervous system, can magnify these effects. Symptoms of overdose in children are excitation, flushed face, dry mouth, fever, hallucinations, and (at higher doses) convulsions.

---

*   Registered trademark of McNeil Consumer Products Co. In Canada: Guelph, ON.
†   Registered trademark of Whitehall-Robins. In Canada: Mississauga, ON.
** Registered trademark of McNeil Consumer Products Co. In Canada: Guelph, ON.

ANTACIDS

Along with analgesics, antacids are among the most widely used nonprescription drugs. While overdoses of antacids generally are not fatal, some adverse effects may result. Antacids contain one or more of the following: sodium bicarbonate, calcium carbonate, magnesium hydroxide, and aluminum hydroxide. Sodium bicarbonate and calcium carbonate are readily absorbed. Large doses of the former may result in sodium retention, whereas large doses of the latter may result in hypercalcemia (increased calcium in the blood).

Magnesium hydroxide and aluminum hydroxide are frequently present in antacids. A common side effect of magnesium hydroxide is diarrhea, whereas the most troublesome side effect of aluminum hydroxide is constipation.

ANTISEPTICS AND ASTRINGENTS

Antiseptics (e.g., iodine and hydrogen peroxide) and astringents (e.g., rubbing alcohol compound) are often found in the medicine cabinet. Iodine is sold as a tincture of iodine in alcohol. Since iodine is corrosive, ingestion causes severe pain in the mouth, throat, and stomach, with nausea, vomiting, and diarrhea occurring. A similar product, mercurochrome, is considered relatively safe.

Hydrogen peroxide may cause skin irritation, even when applied at low concentrations. If concentrated solutions (20 to 30%) of hydrogen peroxide contact the eyes, severe burns and lesions of the cornea may result.

Rubbing alcohol compound or isopropyl alcohol can be used as astringents and disinfectants, or applied to the skin for the relief of muscle aches. Ingestion causes nausea, vomiting, and diarrhea.

MISCELLANEOUS COMPOUNDS

The medicine cabinet contains other compounds that are potentially toxic. One of these is camphor, which is found in high concentration in Vicks Vaporub®*. Ingestion of one gram (= 1000 mg) of camphor (4 teaspoons of Vicks Vaporub®) was found to be lethal in a 1-year-old child.

Syrup of ipecac is a substance that causes vomiting (an emetic) and should be available in all households in case of accidental poisoning, particularly if a medical treatment facility is not readily available. However, no emetic should be used without first obtaining an evaluation of the poisoning from a knowledgeable health care professional (e.g., poison control center staff, physician, or pharmacist). The reason for this is that vomiting may aggravate the symptoms of poisoning in some cases. Acids and alkalis are corrosive (vomiting would cause further damage); strychnine causes convulsions (vomiting would cause choking); and petroleum distillates would cause serious lung problems,

* Registered trademark of Proctor and Gamble Inc. In Canada: Toronto, ON.

if vomiting carried them into the lungs. Therefore, it is necessary to have the poisoning assessed first and also the correct dose of syrup of ipecac recommended.

Chapter **8**

POISONOUS PLANTS

Plants are essential in our environment. They provide shelter for animals, forage for livestock, and food for humans. Plants are such an important part of everyday life that they are cultivated indoors so that a piece of nature can be maintained year-round. What Christmas season would be complete without poinsettia, holly, and mistletoe!

For all their beauty and usefulness, however, plants have an ominous side. Ingestion of plants accounts for approximately 5% of all poisonings, and is exceeded only by ingestion of acetylsalicylic acid (ASA), soaps, detergents, and cleaners. Please note that this chapter covers only *some* of the poisonous plants found in the U.S. and Canada. Some are kept as indoor plants (Table 8.1), whereas others are found outdoors (Tables 8.2, 8.3, 8.4). Only those that cause internal poisoning will be discussed. Plants causing only skin rashes (dermatitis) such as daisies, lady's slipper, juniper, and stinging nettle will not be considered in this book.

Some plants or plant parts are actually used for the treatment of human or animal diseases, in either pharmaceuticals or herbal remedies. The advice of an experienced herbalist should be sought before plants are used medicinally or for making herbal teas. For more information about herbal products, please refer to the book by Hamon and Blackburn (1985) identified in the section, "Further Suggested Reading."

INDOOR PLANTS

Common house plants vary in toxicity from nonpoisonous (such as African violet, Christmas cactus, coleus, and Swedish ivy) to extremely poisonous (such as Jerusalem cherry and oleander). For the most part, ingestion of house plants causes only minor gastric disturbances such as nausea, vomiting, and diarrhea.

Irritation and swelling of the mouth, tongue, and throat are also common effects. This is especially true when plants having calcium oxalate crystals as their toxic agents (such as calla lily, *Dieffenbachia*, elephant ear, and philodendron) are ingested. The irritation is caused by the direct effect of the crystals on the mucous membranes. With some plants (for example, *Dieffenbachia*) the swelling can be so severe that difficulty in breathing and even choking may result.

While most indoor plants (see Table 8.1) cause only minor toxic reactions, there are a few which cause potentially life-threatening effects. Ivy, Jerusalem cherry, and oleander fit into this category. These plants produce toxic agents that result in poisoning when plant parts are swallowed. Death can occur following ingestion of just a small amount of these plants. For example, eating a single leaf of oleander is potentially fatal. The toxicity of the different plant parts often varies. With some plants all parts are equally toxic. With others, the berries, flowers, or roots contain the toxic agents. The colorful parts are especially attractive to children and play a key role in childhood poisonings.

PLANTS IN THE GARDEN

Like indoor plants, cultivated plants, shrubs, and trees found in the garden and yard represent potential sources of toxicity. In contrast to indoor plants, however, a number of outdoor plants produce severe poisoning. Vegetables, such as potato, tomato, and rhubarb, can be toxic if the wrong parts are eaten. Some hedges, shrubs, and trees pose a hazard to livestock.

Toxicity can be produced by accidentally ingesting poisonous plant parts (the leaves, stems, roots, or fruit) or by brewing "teas" from poisonous plants, or by consuming improperly cooked greens. Elderberry branches can be hazardous when used as skewers for barbecuing; the same is true with oleander. For details about garden plants that can produce toxic effects, see Tables 8.2, 8.3, and 8.4.

PLANTS ON THE FARM AND IN THE COUNTRY

Plants present on farms and in the country (such as in fields, wooded areas, and marshes) show the same diversity of toxicity indicated elsewhere. Still, many such plants tend to produce more severe types of toxicity. A few of the poisonous plants on the farm and in the country are listed in Tables 8.5 and 8.6. The book by Humphreys, *Veterinary Toxicology* (written for veterinarians), makes it very clear that there are many additional plants poisonous to livestock. To include all such plants here would exceed the scope of this book.

One concern with "wild" plants is livestock poisoning. Some animals die due to ingestion (unknown to the owner) of poisonous plants. Under normal conditions, livestock avoid these plants. However, under conditions of drought or when the preferred forage is no longer available, livestock tend to turn to such plants. The wise farmer becomes familiar with the poisonous plants in the farm's surroundings and makes every attempt to keep pastures free of them.

Humans are not exempt from the effects of poisonous plants in the field or in the wild. Water hemlock, for instance, has been mistaken for an edible parsnip-like plant, with fatal consequences.

MUSHROOMS

Poisonous mushrooms may grow wherever nontoxic ones do. Some of the most dangerous species are *Amanita phalloides, Amanita verna, Amanita virosa, Gyromitia esculenta,* and the *Galerina* species.

Ingestion of just part of a single mushroom belonging to a dangerous species may be sufficient to cause death. It is wise, therefore, not to pick and eat mushrooms unless they have been identified, very reliably, as being nontoxic. Unfortunately, there are few people who can identify mushrooms as to their edibility. Illustrations and descriptions are usually inadequate as a guide for ensuring that the mushroom is safe.

Most of the mushrooms found growing in "fairy rings" on our lawns are the nontoxic species, *Marasmius oreades*. Although these mushrooms are considered edible by some, it must be realized that they frequently grow together with the poisonous species, *Clitocybe dealbata*. It is, therefore, suggested that unless the picker can distinguish between these two species, mushrooms growing in fairy rings should not be eaten.

TABLE 8.1
Indoor Plants

Plant	Poisonous Part(s)	Toxic Agent(s)	Signs and Symptoms
Amaryllis (*Clivia, Hippaestrum,* or *Lycorus* spp.)	All parts	Lycorin and other alkaloids	Difficulty in swallowing; nausea; vomiting; diarrhea; sweating; trembling
Anthurium (*Anthurium andraenum*)	All parts	Calcium oxalate crystals	Irritation and burning of throat and mouth; swelling of tongue may interfere with breathing and swallowing; choking may occur in severe cases
Azalea (*Rhododendron* spp.)	All parts	Andromedotoxin and its glucosides	Salivation; nausea; vomiting; weakness; difficulty in breathing; incoordination
Calla lily (*Zantedeschia* spp.)	Leaves and rhizome	Calcium oxalate	Burning and swelling of mouth and throat; vomiting
Crown of thorns (*Euphorbia* spp.)	Sap	Unknown irritant	Contact causes skin and eye irritation; ingestion of sap causes swelling of tongue, mouth and throat; vomiting
Cyclamen (*Cyclamen* spp.)	All parts	Cyclamin	Intense stomach cramps; vomiting; diarrhea
Dumbcane (*Dieffenbachia* spp.)	All parts	Calcium oxalate crystals	Irritation and burning of throat and mouth; swelling of tongue may interfere with breathing and swallowing; choking may occur in severe cases
Elephant ear (*Colocasia* spp.)	All parts	Calcium oxalate crystals	Burning and swelling of mouth and throat; salivation; vomiting; diarrhea
Holly (*Ilex* spp.)	Berries	Ilicin glycosides; saponins	Vomiting; diarrhea; stupor

Plant	Toxin	Parts	Symptoms
Hydrangea (*Hydrangea* spp.)	Hycrangin (cyanogenic glycoside)	All parts, especially leaves and buds	Abdominal pain; nausea; vomiting; diarrhea; death
Ivy (*Hedera* spp.)	Hederangenin (saponic glycoside)	Leaves and berries	Excitement; difficulty breathing; coma
Jack-in-the-pulpit (*Arisaema triphyllium*)	Calcium oxalate crystals	All parts, especially rhizome	Burning and intense irritation of mouth and throat
Jerusalem cherry (*Solanum pseudocapsicum*)	Solanine	Leaves and unripe fruit	Stomach pains; low body temperature; paralysis; dilated pupils; vomiting; diarrhea; circulatory and respiratory depression; loss of sensation; death
Mistletoe (*Phoradendron* spp.)	Toxic amines	All parts, especially berries	Acute stomach and intestinal irritation; diarrhea; slow pulse; decreased heartbeat
Oleander (*Nerium oleander*) (Notice: A warning to travellers in tropical areas: even smoke from oleander branches can poison you)	Nerioside; oleandroside (cardiac glycosides)	Twigs, green or dry leaves, and flowers	Nausea; severe vomiting; stomach pain; dizziness; slowed pulse; irregular heartbeat; dilated pupils; bloody diarrhea; drowsiness; unconsciousness; paralysis of respiration; death
Philodendron (*Philodendron* spp.)	Calcium oxalate crystals	Leaves and stems	Burning of mouth; vomiting; diarrhea
Poinsettia (*Euphorbia pulcherrima*)	Euphorbin	All parts, particularly sap (Note: toxicity recently disputed)	Abdominal pain; vomiting; diarrhea
Rhododendron (*Rhododendron* spp.)	Andromedotoxin: arbutin glycoside	All parts	Salivation; nausea; vomiting; dizziness; difficulty in breathing; lack of coordination
Rubber plant (*Ficus* spp.)	Ficin; furocoumarin; ficusin; psoralene	Sap	Vomiting; diarrhea; exhaustion; severe irritation of gastrointestinal tract; damage to skin (rash, dermatitis, blistering)

TABLE 8.2
Flowering Plants in the Garden

Plant	Poisonous Part(s)	Toxic Agent(s)	Signs and Symptoms
Autumn crocus (*Colchicum autumnale*)	All parts, especially bulbs	Colchicine	Irregular heartbeat; confusion; burning sensation; nausea; diarrhea; weakness; death
Bleeding heart (*Dicentra* spp.)	All parts	Protopine	Trembling; staggering; weakness; difficulty in breathing; convulsions
Calla lily (*Zantedeschia* spp.)	Leaves and rhizome	Calcium oxalate crystals	Irritation of mouth; vomiting
Castor bean (*Ricinus communis*)	Seeds	Ricin	Handling leaves and seeds may cause itching rash and broken blisters; chewing seeds causes burning in mouth, throat, and stomach; loss of appetite; abdominal pain; vomiting; diarrhea; excessive thirst; dullness of vision; convulsions; death
Daffodil, narcissus (*Narcissus* spp.)	All parts, especially bulbs	Alkaloid	Nausea; vomiting; diarrhea; trembling; convulsions; may be fatal
Delphinium, larkspur (*Delphinium* spp.)	All parts, especially young plants and seeds	Diterpenoid alkaloids	Nausea; abdominal cramps; bloating; twitching muscles; paralysis; death
Foxglove (*Digitalis* spp.)	All parts	Cardiac glycosides	Nausea; diarrhea; abdominal pain; headache; confusion; blurred vision; irregular heartbeat; convulsions; death
Hyacinth (*Hyacinthus orientalis*)	All parts, especially bulbs	Alkaloid	Intense stomach cramps; nausea; vomiting; diarrhea

Plant	Toxin	Toxic parts	Symptoms
Iris (*Iris* spp.)	Irisin	Leaves and rhizomes	Nausea; vomiting; diarrhea
Lily of the valley (*Convallaria majalis*)	Convallarin; convallamarin glycosides	All parts	Irregular heartbeat; nausea; confusion; circulatory collapse; death
Lobelia (*Lobelia* spp.)	Lobelamine; lobeline	All parts	Nausea; progressive vomiting; exhaustion and weakness; tremors; convulsions; coma; death
Lupins (*Lupinus* spp.)	Quinolizidine alkaloids; piperidine alkaloids	All parts	Nausea; diarrhea; abdominal pain; blurred vision; irregular heartbeat; convulsions; death
Monkshood (*Aconitum* spp.)	Aconitine; other alkaloids	All parts	Restlessness; salivation; nausea; impaired vision; death
Morning glory (*Ipomoea* spp.)	Ergot alkaloids	Seeds	Nausea; digestive upset; blurred vision; mental confusion; coma; hallucinogenic reaction when 50 or more seeds are ingested
Poppy (*Papaver* spp.)	Alkaloids	All parts, especially raw (except edible poppy seeds)	Eating unripe fruit: deep sleep; dizziness; delirium; slow breathing; death. Eating other parts: nausea; vomiting; stomach pains; twitching muscles
Star-of-Bethlehem, (*Ornithogalum umbellatum*)	Cardiac glycosides	All parts	Depression; salivation; vomiting; diarrhea; labored breathing; rapid pulse; bloody urine; death
Sweet pea (*Lathyrus odoratus*)	Beta-(gamma-L-glutamyl)-aminopropionitrile	Seeds	Prickly sensation; cramps; lameness; paralysis; slow and weak pulse; shallow breathing; convulsions; death
Tulip (*Tulipa* spp.)	Tulipene	Bulb	Vomiting; diarrhea

TABLE 8.3
Vegetables

Plant	Poisonous Part(s)	Toxic Agent(s)	Signs and Symptoms
Potato (*Solanum tuberosum*)	Leaves, vines, sprouts, and *green*-skinned potatoes	Solanine; chaconine	Digestive upset; cold perspiration; lowered temperature; dilated pupils; confusion; weakness; numbness; paralysis; death
Rhubarb (*Rheum rhaponticum*)	Leaves	Oxalic acid; soluble oxalates	Stomach pains; nausea; vomiting; weakness; difficulty in breathing; burning of mouth and throat; internal bleeding; coma; death
Tomato (*Lycopersicon esculentum*)	Flowers, leaves, stems, root tips, and buds	Tomatine; solanine; oxalic acid	Nausea; vomiting; abdominal pains; constipation or bloody diarrhea; sluggishness; salivation; difficulty in breathing; reduced heartbeat; trembling; weakness; loss of feeling; paralysis; death

TABLE 8.4
Hedges, Shrubs, and Trees

Plant	Poisonous Part(s)	Toxic Agent(s)	Signs and Symptoms
Apple (*Malus* spp.)	Foliage and seeds	Amygdalin (cyanogenic glycoside)	Abdominal pain; nausea; vomiting; difficulty in breathing; twitching; spasms; coma; death
Bittersweet (American) (*Celastrus* spp.)	All parts	Alkaloids	Nausea; vomiting; diarrhea; exhaustion; weakness; coma; convulsions
Bittersweet (European), climbing nightshade (*Solanum dulcamara*)	All parts	Solanine; solanicine	Burning sensation; nausea; dizziness; dilation of pupils; vomiting; abdominal pain; diarrhea; depression; shock; difficulty breathing
Burning bush (*Euonymus* spp.)	Leaves, bark, and fruit	Unknown	Vomiting; diarrhea; weakness; chills; coma; convulsions
Chokecherry* (*Prunus virginiana*)	All parts; fruit is safe if pits are removed	Amygdalin (cyanogenic glycoside)	Abdominal pain; nausea; vomiting; death
Daphne (*Daphne* spp.)	All parts (but not all species; general caution advisable)	Dihydroxy-coumarin-type glycosides	Blisters on skin; burning or ulceration of mouth, throat, and stomach; internal bleeding with bloody diarrhea; convulsions; coma; death
Elderberry (or elder) (*Sambucus* spp.)	All parts (but not all species); includes roots and especially unripe berries	Cyanogenic glycoside	Nausea; vomiting; diarrhea
Honeysuckle (*Lonicera* spp.)	Berries — possibly (depends on species)	Unidentified	Nausea; vomiting; diarrhea; irregular heartbeat; sensitivity to sunlight; coma; shock-like symptoms
Hydrangea (*Hydrangea* spp.)	All parts, especially leaves and seeds	Hydrangin (cyanogenic glycoside)	Abdominal pain; nausea; vomiting; diarrhea; death
Lantana (*Lantana* spp.)	Green, unripened berries	Lantanin alkaloid; lantadene A	Stomach and intestinal irritation; muscular weakness; circulatory collapse; death. Acute symptoms resemble atropine poisoning

TABLE 8.4 (continued)
Hedges, Shrubs, and Trees

Plant	Poisonous Part(s)	Toxic Agent(s)	Signs and Symptoms
Oak* (*Quercus* spp.)	Leaves, unleached acorns, and young shoots	Tannic acid	Loss of appetite; constipation; abdominal pain; excessive thirst; frequent urination; bloody diarrhea; weak pulse; death
Sedum (*Sedum acre*)	All parts	Unidentified glycosides	Vomiting; diarrhea; weakness; respiratory depression
Virginia creeper (*Parthenocissus quinquefolia*)	Berries	Unknown	Evidence is circumstantial, but it is believed that ingestion of berries has caused death of children
Yew* (*Taxus* spp.)	All parts	Taxine	Vomiting; diarrhea; trembling; difficulty in breathing; muscle weakness; slow heartbeat; convulsions; coma; death

Note: * indicates livestock hazard.

TABLE 8.5
Plants on the Farm

Plant	Poisonous Part(s)	Toxic Agent(s)	Signs and Symptoms
Arrowgrass* (*Triglochin maritima, T. palustris*)	All parts	Cyanogenic glycoside	Rapid or deep breathing; muscular spasms; convulsions; respiratory paralysis; death
Black nightshade, deadly nightshade [Bittersweet] (*Solanum nigrum, S. dulcamara*)	All parts, especially unripened fruit	Solanine; solanidine	Headache; stomach pain; low body temperature; paralysis; dilated pupils; vomiting; diarrhea; shock; circulatory and respiratory depression; death
Bracken fern* (*Pteridium aquilinum*)	All parts, green or dry, especially rhizome	Thiaminase	Incoordination; lethargy; muscular twitches; tremors; convulsions; death
Buttercup (*Ranunculus* spp.)	All parts, except seeds	Protoanemonin	Severe skin irritation; abdominal pain; diarrhea; increased salivation
Coneflower, Black-eyed Susan (*Rudbeckia* spp.)	All parts	Unknown	Abdominal pain; incoordination; rapid breathing
Greasewood* (*Sarcobatus vermiculatus*)	All parts, especially buds and young leaves	Soluble oxalates	Depression; weakness; weak pulse; shallow breathing; collapse
Groundsel* (*Senecio* spp.)	All parts	Pyrrolizidine alkaloids	Abdominal pain; nausea; vomiting; enlarged liver; headache; apathy; emaciation
Jimsonweed (*Datura stramonium*)	All parts, especially seeds and leaves	Hyoscyamine; atropine	Thirst; dilated pupils; dry mouth; redness of skin; headache; hallucinations; nausea; rapid pulse; elevated body temperature; increased blood pressure; delirium; convulsions; coma; death
Locoweed*, milkvetches* (*Oxytropis and Astragalus* spp.)	All parts	Locoine	Incoordination; difficulty in breathing; nasal discharge; frequent urination; inability to eat or drink; paralysis; death
Lupins* (*Lupinus* spp.)	All parts, especially seeds	Quinolizidine alkaloids; piperidine alkaloids	Difficulty in breathing; twitching; convulsions; unconsciousness; death
Milkweed (*Asclepias* spp.)	All parts	Calitoxin	Severe upset of stomach and intestine

TABLE 8.5 (continued)
Plants on the Farm

Plant	Poisonous Part(s)	Toxic Agent(s)	Signs and Symptoms
Monkshood* (*Aconitum* spp.)	All parts, especially roots and seeds	Aconitine; other alkaloids	Restlessness; salivation; weakness; irregular heartbeat; nausea; dizziness; anxiety; impaired speech and vision; death
Pokeweed (*Phytolacca americana*)	All parts, especially rootstock	Unknown	Severe stomach cramps and pain; nausea with persistent vomiting; diarrhea; difficulty in breathing; weakness; spasms; severe convulsions; death
Tansy (*Tanacetum vulgare*)	All parts	Tanacetin	Convulsions; violent spasms; dilated pupils; rapid and weak pulse; death

Note: * indicates livestock hazard.

TABLE 8.6
Plants in Wooded and Marsh Areas

Plant	Poisonous Part(s)	Toxic Agent(s)	Signs and Symptoms
Baneberry (*Actaea* spp.)	All parts, especially berries and root	Essential oil	Acute stomach cramps; headache; rapid pulse; vomiting; delirium; dizziness; circulatory failure; death
Common cattail* (*Typha latifolia*)	Leaves and stem	Unknown	Stiffness; profuse perspiration; tremors
Death camas* (*Zigadenus* spp.)	All parts, especially bulbs	Steroid alkaloids	Abdominal pain; nausea; vomiting; increased salivation; muscular weakness; difficulty in breathing; lowered body temperature; coma; death
Horsetails* (*Equisetum* spp.)	All parts	Possibly thiaminase	Muscular weakness and exhaustion; ataxia; rapid and weak pulse; prostration; coma; death
Mountain laurel* (*Kalmia polifolia* var. *microphylla*)	All parts	Andromedotoxin	Watering of mouth, eyes, and nose; slow pulse; lowered blood pressure; incoordination; convulsions; paralysis; coma; death
Poison hemlock (*Conium maculatum*)	All parts, especially seeds and root	Coniine; other alkaloids	Vomiting; diarrhea; muscular weakness; paralysis; nervousness; trembling; dilated pupils; weak pulse; convulsions; coma; death
Poison ivy, poison oak, poison sumac (*Toxicodendron* spp.)	Sap	Urushiol	Burning, itching rash; may lead to large blisters; local swelling and fever
Skunk cabbage (*Symplocarpus foetidus*)	All parts	Calcium oxalate crystals	Burning and intense irritation of mouth and throat
Water hemlock* (*Cicuta* spp.)	Leaves and root	Cicutoxin	Stomach pain; nausea; vomiting; diarrhea; elevated body temperature; dilated pupils; difficulty in breathing; rapid and weak pulse; tremors; delirium; violent convulsions; death

Note: * indicates livestock hazard.

Chapter 9

PAINTS, SOLVENTS, CLEANSING AGENTS, AND ALL KINDS OF THINGS

The home, workshop, and garage are places where highly toxic substances can be found. Compounds such as paint removers, paintbrush cleaners, grease solvents, antifreeze, deicers, and glue are readily available and are potentially harmful when used improperly or carelessly. All these products are effective partly because of their organic solvent content. Unfortunately, these solvents are unable to distinguish the grease on the garage floor from human fat tissue and body membranes. When swallowed, these solvents act directly on the membranes to produce marked irritation. As indicated in Table 9.1, there are a number of products that are potentially quite toxic and that will be described in greater detail.

TABLE 9.1
Toxic Components of Paints, Solvents, and Related Products

Products	Toxic Components
Paintbrush cleaners	Acetone; caustic alkalis; methanol; turpentine
Paint	Hydrocarbons; petroleum distillates; lead
Paint removers and solvents	Alcohols: amyl, butyl, ethyl, methyl; benzene; carbon tetrachloride (in older preparations); caustic alkalis; kerosene; toluene
Antifreeze/deicers	Alcohols: isopropyl, methyl; ethylene glycol
Glue	Toluene; xylene

PAINTS

Latex and oil-based paints are the two common types of paint. Latex paints have a water base and contain elastomers, titanium oxide, zinc oxide, ethylene glycol, and inert filler pigments. Some latex paints may also include mercury-based fungicides as a preservative. Oil paints have linseed oil and alkyd resin or oil varnish as a base, and contain other chemicals similar to those found in latex paints.

The toxicity of latex paints is low, even if the product contains the solvent, ethylene glycol. However, irritation of skin, eyes, and mucous membranes can occur. Such symptoms are also reported after the use of oil-based paints, which contain petroleum products. Oil-based paints can also cause headache, dizziness, and nausea. The risk of such symptoms is increased if ventilation is poor, e.g., when painting indoors during cold weather, when doors and windows are closed.

Some exterior oil-based paints contain lead salts (up to 1%). In the past, lead concentrations were as high as 40%. Many older homes, built before 1955, have leaded paint both indoors and outdoors. Renovating these homes has resulted in high levels of paint particles, with lead being dispersed into the home environment. Unless special precautions are taken, the risk from the process of removing leaded paint can exceed the risk from leaving it as is. Lead poisoning can occur when small children, either from boredom or out of curiosity, eat flaking paint particles or chew on toys or furniture painted with lead-containing products.

Lead primarily affects the gastrointestinal tract and the brain. Gastrointestinal effects include loss of appetite, vomiting, abdominal pain, and constipation. Effects on the brain include irritability, drowsiness, incoordination, convulsions, and coma. In children, learning disabilities, behavioral problems, decreased I.Q., and decreased growth may result from early childhood exposure to lead.

PAINT CLEANERS, REMOVERS, STRIPPERS, SOLVENTS, AND THINNERS

Paint (and paintbrush) cleaners, removers, strippers, thinners, and solvents are widely used. They contain a variety of different chemicals. As an example, paintbrush cleaners usually contain acetone and turpentine and sometimes methanol. Acetone has an aromatic odor and pungent taste. Ingestion of small amounts of acetone results in nausea, vomiting, and diarrhea. If large amounts are consumed, coma (state of unconsciousness from which individuals cannot be aroused) and death may result. Inhalation of acetone vapors produces coughing, lung irritation, headache, and fatigue. For a brief discussion of the toxicity of methanol, see the section "Antifreeze and Deicers" (below).

Turpentine is obtained from the distillation of pine wood. Ingestion of turpentine leads to abdominal pain, nausea, vomiting, and diarrhea. Subsequently, a weak, rapid pulse develops along with reduction of central nervous system function and respiratory failure. Prolonged exposure to turpentine vapors causes dizziness, inflammation of the eyes and nasal passages, bronchitis, pneumonia, rapid heartbeat, and rapid breathing.

Paint removers and solvents are used to remove paint, wax, lacquers, and grease. Chemicals such as kerosene, benzene, toluene, gasoline, and, formerly, carbon tetrachloride, are used as solvents. In general, these solvents produce effects similar to those observed with paintbrush cleaners. Solvents are used both as paint thinners and as drain cleaners for grease removal. The most common paint stripper is methylene chloride. Methylene chloride may irritate skin or eyes and cause headache, lethargy, nausea, and dizziness. Although methylene chloride has been shown to cause cancer in mice, evidence of human carcinogenicity is inadequate. Lacquer thinners are composed of aromatic hydrocarbon solvents, petroleum distillates, aliphatic alcohols, acetates, and ketones.

Water-soluble strippers based on less familiar specialty products (e.g., N-methyl-pyrrolidone and butyrolactone) are now available and have considerable appeal as "environmentally friendly" alternatives. While these solvents do not have the familiar "solvent" smells, pyrrolidones readily penetrate the skin, and can also carry other chemicals into the skin with them. Butyrolactone ingestion can lead to depressed brain function and coma.

Petroleum distillates, kerosene, gasoline, ethanol, and methanol can cause central nervous system depression, disorientation, and tissue damage. Aspiration (inhaling vomited material into the lung) of petroleum distillates can cause hydrocarbon or aspiration pneumonia. Ingestion of kerosene causes nausea, vomiting, coughing, and irritation to the lungs.

Benzene and toluene are volatile hydrocarbons used as industrial solvents. Exposure is primarily by inhalation of the vapors, although absorption through the skin may also occur. These solvents affect the respiratory and gastrointestinal tracts. Intentional toluene sniffing (see later) results in damage to the kidneys and brain. Chronic exposure of workers to high doses of benzene has caused cancer of the bone marrow.

Gasoline is a petroleum product composed of a variety of hydrocarbons. Although it is not intended for use as a solvent, it is often used for this purpose. Toxic effects are similar to those produced by kerosene. A greater hazard exists from gasoline vapors. At low concentrations, inhaling gasoline vapors causes flushing of the face, staggering, mental confusion, disorientation, slurred speech, and difficulty swallowing. At high concentrations, loss of consciousness, coma, and death may occur. Of course high concentrations of gasoline vapors can also be an explosion hazard.

ANTIFREEZE AND DEICERS

Antifreeze and deicers are typically composed of methanol (also called methyl alcohol or "wood alcohol") or ethylene glycol. Methanol is also used as a paint remover and a solvent in shellac and varnish. Symptoms are observed following ingestion of the liquid or inhalation of the fumes. Unfortunately, methanol can be mistaken for ethanol (grain alcohol). Ingestion of methanol causes serious injury. Initial effects include severe gastrointestinal cramps with vomiting and blurred vision. Permanent blindness and irreversible brain damage also may result from ingestion of methanol.

Ethylene glycol is often responsible for the poisoning of children and pets. This is due primarily to the pleasant (sweet) taste of ethylene glycol. Symptoms of poisoning include vomiting, extreme weakness, kidney problems, unconsciousness, and convulsions.

GLUE

Commercially available glues often contain organic solvents such as toluene and xylene. Inhaling the fumes can cause giddiness, headache, dizziness, confusion, stupor, and coma. Death may result from respiratory failure or sudden heart failure. Swallowing such solvents causes burning in the mouth and throat, hoarseness, nausea, vomiting, salivation, and coughing. Contact with the skin causes redness and blisters.

CLEANSING AGENTS

Common household products are by no means free of toxic effects. Many household cleaners (such as detergents, bleaches, and caustic acids and alkalis) can be as hazardous as the solvents used in industry. Careless use and disposal can lead to tragic results. Table 9.2 lists common household products and some of their toxic components.

TABLE 9.2
Some Household Products and Their Toxic Components

Product	Toxic Components
Bleaches	Sodium hypochlorite; alkaline borates
Cleaners	Soaps; alkaline borates; detergents: anionic, nonionic; polyphosphates; glycols
Disinfectants	Detergents: anionic, nonionic, cationic; phenol; isopropyl alcohol; pine oil; petroleum distillates
Deodorants	Detergents: anionic, nonionic; ethanol; soap
Deodorizers	Chlorinated hydrocarbons; insecticides; detergents: anionic, nonionic; hydrocarbons; petroleum distillates
Drain cleaners	Sodium hydroxide; trichloroethane; sodium hypochlorite; surfactants
Liquid polishes and waxes	Hydrocarbons; petroleum distillates; isopropyl alcohol; borates; xylene; toluene
Mothballs	Naphthalene; paradichlorobenzene; chlorinated hydrocarbons
Perfumes	Ethanol; essential oils
Shampoos	Detergents: anionic, nonionic, cationic; soaps

DETERGENTS

Common cleansing agents include laundry detergents, automatic dishwasher detergents, and hand dishwashing detergents, all of which may cause toxic effects. Of all accidental poisonings in children less than 5 years old, 6% are due to the ingestion of detergents. There are three classes of detergents:

anionic, nonionic, and cationic. They differ in their toxicity, with anionic and nonionic detergents being less toxic than cationic ones.

Anionic detergents are present in handwashing liquids, dishwasher granules and powders, shampoos, and deodorants. Ingestion causes diarrhea, intestinal distention, and occasionally vomiting. All these detergents can irritate the skin (especially after prolonged exposure) by removing the natural oils, which may result in redness and soreness. In sensitive people, thickening of the skin with cracking, scaling, and blistering can occur. Nonionic detergents, often found in hand dishwashing liquids, shampoos, laundry detergents, and whiteners, are relatively nontoxic. They are considered harmless by ingestion and are only slightly irritating to the skin. Cationic detergents are used to destroy bacteria on cooking equipment, sickroom supplies, and diapers, and are used in fabric softeners. Cationic detergents are corrosive to the skin, eyes, and mucous membranes.

BLEACH

Bleaches are solutions of sodium hypochlorite, sodium perborate mixes, or chloroisocyanurate mixes. Such compounds are found in household and commercial laundry bleaches as well as hair bleaches. Household bleach is usually 5% sodium hypochlorite. It causes mild to moderate skin and eye irritation. The ingestion of bleach accounts for 5% of all poisonings in children under 5 years of age. Bleach is harsh to the mucous membranes and causes irritation and burning of the mouth, throat, and stomach. Pain and vomiting may result.

CORROSIVE ACIDS AND ALKALIS

Corrosive acids (such as hydrochloric acid, phosphoric acid, or sulfuric acid) and alkalis (such as ammonia or sodium hydroxide [lye]), are common components of household cleaners. These highly toxic agents are present in cleaners for jewellery, windows, and floors (ammonia); drain and oven cleaners (sodium hydroxide, sodium carbonate, or ammonia); toilet bowl cleaners (hydrochloric acid or phosphoric acid); and metal cleaners (hydrochloric acid, sulfuric acid, or oxalic acid). Concentrations of these compounds can be high. For example, granulated drain cleaners can contain up to 50% sodium hydroxide, whereas liquid drain cleaners usually contain 8.5% sodium hydroxide.

Corrosive acids and alkalis are toxic by all routes of exposure. Inhalation of vapors causes inflammation of the respiratory passages, coughing, chest pain, and difficulty in breathing. The effects of ingestion include burns to the mouth, throat, and stomach and severe pain in the mouth, chest, and abdomen. Scarring of the gastrointestinal tract can lead to long-term disability. Skin contact results in burns, pain, and brownish or yellowish stains. Splashing corrosive agents in the eyes causes pain, tearing, sensitivity to light, and possible destruction of the cornea. The extent and severity of the injuries depends on the type of acid or alkali, the quantity, concentration, length of contact, and the presence or absence of food.

ALL-PURPOSE CLEANERS AND POLISHES

Household washing and general purpose cleaners can contain the synthetic detergents and some of the alkalis mentioned above. They may also contain pine oil (see "Disinfectants," below). These cleaners can cause skin and eye irritation or nausea and vomiting if swallowed. Furniture polishes containing lemon oil are very dangerous if ingested, due to the high risk of inhaling vomited material into the lungs, resulting in pneumonia.

GLASS CLEANERS

Glass cleaners generally contain isopropyl alcohol and cellosolves (glycol ether solvents), synthetic anionic detergents, small amounts of ammonia, and water. The toxicity of these products is relatively low unless a considerable amount of alcohol is present. Isopropyl alcohol ingestion can cause nausea, vomiting, and abdominal pain. The cellosolves can penetrate intact skin, reduce brain and nervous system activity, and can cause kidney injury if large amounts are swallowed. Household glass cleaners do not contain methanol but the more toxic industrial glass cleaners do.

RUG CLEANERS AND RUG DEODORANTS

Many different formulations of rug cleaners and deodorants are available; some contain synthetic detergents with alcohol, and water; some contain soda soap, alkaline water softening agents, borax, naphthalene, essential oils, and combinations of soap, solvent, and cellosolve. The powders may contain washing soda. These formulations are generally of low to moderate toxicity. Some spot cleaners for rugs and upholstery can contain methylene chloride or perchloroethylene. Such formulations can be toxic with overexposure.

DISINFECTANTS

Most disinfectants contain a combination of substances, e.g., detergents, pine oil, phenol, isopropyl alcohol, cresol, and petroleum distillates. Pine oil is chemically related to turpentine. It produces nausea, vomiting, pain, and diarrhea as well as irritation of the eyes. Phenol causes nausea, vomiting, collapse, and coma, and it corrodes the skin. Isopropyl alcohol (isopropanol) ingestion causes nausea, vomiting, and abdominal pain.

Cresol is a phenol derivative. It is used as an antiseptic, disinfectant, and germicide. If applied to the skin, cresol causes tissue destruction and scar tissue later. Ingestion results in toxic effects throughout the body. Initial effects are in the gastrointestinal tract with pain, nausea, vomiting, and diarrhea. Later, cardio-vascular collapse, breathing difficulty, and severe kidney damage may occur.

DANGEROUS MIXTURES

Most chemical products used in the home are designed to be used alone, either at full strength or diluted with water according to label directions. It is natural to assume that if ammonia or vinegar is not cleaning something very well, the addition of a little bleach might help. This assumption is dangerous.

Chlorine bleaches, composed of hypochlorites, react with acids (in vinegar, drain cleaners, toilet bowl cleaners, or rust cleaners) to form chlorine gas, or with ammonia to form chloramine gas. Both gases are highly toxic.

MOTHBALLS

Although they appear innocuous, mothballs can be quite toxic. Mothballs contain either naphthalene or paradichlorobenzene. Naphthalene is the more toxic ingredient. Consumption of such mothballs results in fever, pallor, lethargy, abdominal pain, diarrhea, loss of appetite, vomiting, and headache. An additional effect of naphthalene ingestion is a severe and rapid destruction of red blood cells. Inhalation of naphthalene vapors produces headache, mental confusion, and blurred vision.

Another hazard is the residue of naphthalene which remains on clothes stored directly in contact with mothballs. Naphthalene is not soluble in water, so washing will not remove it from baby clothes. Baby oil acts as a solvent for the residues so that the naphthalene can be rapidly absorbed through the child's skin.

Paradichlorobenzene (technically, *para*-dichlorobenzene) is the ingredient more commonly found in a newer type of mothball. Although it is less toxic than naphthalene, it is irritating to skin, eyes, and mucous membranes. Inhaling the vapors can result in headache and dizziness. Ingestion causes nausea, vomiting, and diarrhea. Paradichlorobenzene has been shown to cause cancer in animals, but evidence for human carcinogenicity is inadequate.

COSMETICS

In general, cosmetics have a low order of toxicity since large quantities (greater than 100 g) would have to be consumed to produce life-threatening effects. However, cosmetics such as cologne or aftershave lotion may contain 50 to 80% alcohol and could be harmful if swallowed by a child. Other problems with cosmetics are skin sensitization and allergic reactions. Cosmetics that have been linked to adverse health effects are hair dyes, hair sprays, hair-removal agents, and nail polish remover.

Hair dyes vary in toxicity. Those utilizing vegetable-based dyes are considered safe. In contrast, some permanent hair dyes contain toxic metallic preparations based on cobalt, copper, cadmium, iron, lead, nickel, silver, bismuth, or tin. These can be harmful. Others contain synthetic organic dyes such as paraphenylenediamine, a strong sensitizing (allergy-producing) agent which can lead to permanent blindness if it comes in contact with the eyes. Most permanent dyes also contain hydrogen peroxide 6%, which is a weak irritant and is of low toxicity. There has been a concern that dark hair dyes, used for many years, might cause one type of cancer (non-Hodgkin's lymphoma; NHL). At this time, only long-term use of black dyes appears to be related to NHL, but more studies are needed.

Hair sprays contain natural and synthetic resins. If these are inhaled, respiratory problems can occur. This is more likely to happen to hair stylists, since they are chronically exposed to hair sprays.

Depilatory (hair-removal) agents usually contain soluble sulfides or calcium thioglycolate. They may irritate the skin and, if ingested, the gastrointestinal tract. Ingestion of large doses results in low blood sugar, convulsions, and respiratory failure.

Acetone is the main component of nail polish remover. Acetone is also present in varnishes, glues, and nail polish. It is a solvent which, if swallowed, can produce nausea, vomiting, and diarrhea.

Chapter 10

INDUSTRIAL CHEMICALS

INDUSTRIAL vs. NATURAL CHEMICALS

A common misconception that should be identified and overcome before people can sensibly discuss chemicals is that chemicals made by nature are good, whereas those made by people are bad. This is certainly not the case. Quite often nature and industry produce the same chemical; nature produces the most toxic of all chemicals; and the "goodness" or "badness" of a chemical often has much more to do with how it is managed or used than with its own properties. Certainly it is a mistake to fear "chemicals" as a general group: we ourselves are made up of chemicals. Sugar is a very familiar food, but its full chemical name sounds very strange. To the specialist, ordinary table sugar is alpha-D-glucopyranosyl-beta-D-fructofuranoside; strange sounding chemical names do not have to indicate a risk. Natural foods are 100% chemicals; the smell of apple pie (and the pie itself, not to mention its dish) is made of chemicals. All people are totally surrounded by chemicals, even in the pristine wilderness, so a general fear of chemicals is not very reasonable. To some extent, the good judgment of the general public has been swayed toward fear of chemicals by a few sensationalists through the media and in a minority of environmental groups.

What *is* special about industrial chemicals? Many are natural products that are refined and moved to a different location; many others are not found in nature, but are made synthetically in large amounts because they offer some advantage — some greater benefit — compared to what is available naturally. Thus, chemicals are everywhere, are not necessarily "bad," and provide us with many benefits.

BALANCING THE BENEFITS AND THE RISKS

Industrial chemicals are essential to the way of life widely appreciated, and generally taken for granted, in the U.S. and Canada today. These chemicals are used to disinfect, to purify drinking water, in making all sorts of electronic devices (including computers and TVs), and in making paints, glues, dyes, toners in printers and photocopiers, plastics, fabrics, space satellites and shuttles, fuels, and much more. They are used as intermediates in the production of drugs used to combat diseases of people and livestock, and in sunscreens to block harmful ultraviolet (UV) rays. These are material benefits. There are also economic benefits: the industrial chemical industry in the U.S. and Canada amounts to more than a quarter of a trillion dollars each year. Some of this is private economic gain (for owners and investors), while much of it is public economic gain (employment, corporate taxes). It is evident that we experience many benefits, of different kinds, from the production and use of industrial chemicals.

Nonetheless, the manufacturing, distribution, and use of industrial chemicals, usually on a large scale, poses particular risks. First, there is the chance that some accident may occur during production. Second, some workers are exposed to the chemicals during all their working hours. Third, environmental hazards exist when the chemicals are distributed and used.

CRADLE-TO-GRAVE OR LIFE CYCLE MANAGEMENT OF CHEMICALS

The chemical industry has learned much from the accidents of the past two decades, and has introduced now widely practiced "Responsible Care" and "Product Stewardship" programs. The cradle-to-grave management of industrial chemicals involves continuous review of each chemical, from its conception in the research laboratory, through applied research and development, pilot testing, scale-up, plant construction and operation, distribution, marketing, uses, and the handling of waste and of emptied containers and packaging. The life cycle approach to comprehensive management of chemicals is basically the same as cradle-to-grave, except that the emphasis is not upon the "grave," which suggests burial, but rather on "cycle," which suggests reuse and recycling the waste. The chemical industry is developing many new means of distributing its products, including special pesticide containers which are drained into farm equipment without any exposure of the operator, returnable and rechargeable containers, and optimized washing and recycling or high temperature incineration of emptied chemical containers.

Such management programs are endorsed and used both by the chemical industry and by government agencies in most industrialized countries. These programs have done much to reduce the risks associated with the manufacture, marketing, and disposal of chemicals. Part of the responsibility for chemical safety remains with the individual user of the chemical.

PREDICTING ENVIRONMENTAL IMPACTS OF INDUSTRY

Industrial processes or plants need to be designed and operated in such a way as to maintain both the health of people (directly or indirectly involved in the work, those nearby, and users of the end products) and the health of the environment. Toxicology and ecotoxicology (see later section "Ecotoxicology to Assess Risks of Polluting the Environment") are important components in assessing the impact a particular process or plant may have on human health and ecosystem health. Health risks to the workers are generally examined by specialists in occupational health and safety, but the possible risks to health of nearby residents and users are assessed along with potential effects upon the rest of the ecosystem; these are addressed by teams of engineers, computer modelers, toxicologists and ecotoxicologists, and others such as archeologists, economists, etc. This latter type of assessment is called Environmental Impact Assessment (EIA). From a toxicological and ecotoxicological point of view, the task is: to determine the total hazard (how much, of which chemicals, of what toxicity, where); to determine what pathways exist for these chemicals to escape from the process or plant and to reach an average or particularly sensitive individual person, habitat, or species of animal or plant.

For each identified pathway (e.g., surface water flow) the rate at which each toxic pollutant is delivered or accumulated is calculated. The predicted concentrations of each chemical arriving via *all* pathways are then plotted on a map of the area. The susceptibilities of each species in the area (including humans) are then compared with the predicted concentration at each pertinent location to give the incremental (i.e., added) risk. Finally, the incremental risk is compared with risk values that the public typically considers acceptable — usually a one in 100,000 or one in 1,000,000 chance of experiencing a significant health problem in a lifetime. This kind of assessment forms a major chapter in the report (Environmental Impact Statement or EIS) arising from the EIA process. The EIS is usually submitted to both technical and public review, and the plans are accepted, returned for modification, or rejected.

SOME EXAMPLES OF ENVIRONMENTAL POLLUTANTS

Although the production of chemicals is carried out in such ways that accidents are unlikely to occur, disastrous accidents have taken place, and have received wide publicity, such as the 1976 incident in Seveso, Italy, where particularly toxic dioxins were released, or the 1984 incident in Bhopal, India, where about 2500 people were killed by methyl isocyanate. As terrible and regrettable as these incidents are, they are really exceptions, considering the large number of chemical plants and the vast amount of chemicals produced. This is due to the fact that engineers and chemists constantly improve the facilities and keep a close watch on possible trouble spots. Numerous regulations and guidelines determine what protective equipment workers in these plants have to use and describe the occupational exposure limits to the chemicals.

Since these accidents, both government and industry initiatives have led to better labeling of pipes, tanks, and valves in chemical plants, as well as to better systems for informing surrounding neighborhoods about safety measures and emergency procedures. As well, it has become mandatory for employers to provide workers with both information and training to minimize risks to their health when handling chemicals (see section "Safe Use of Industrial Chemicals").

Once chemicals have left the plant and the distribution system and are used another problem arises: release of the chemical into the environment and exposure of larger or smaller groups in the population. It is impossible to give many examples in this book, but a few highly publicized ones will be discussed briefly.

PBBs

A commercial flame retardant, consisting of a mixture of polybrominated biphenyls (PBBs) was accidentally mixed into dairy cows' feed in Michigan in 1973. Milk production and feed consumption dropped within 3 weeks, and eventually nearly 30,000 cattle had to be destroyed. However, by that time, the chemical had been distributed widely among other livestock species and had been found in humans. Scientists are still wondering what long-term effects the PBBs will have. Many people with PBBs in their bodies complain of excessive fatigue, aching bones and muscles, and other symptoms.

PCBs

Another chemical family that has made the headlines more recently is the polychlorinated biphenyls (PCBs). PCBs have been used since 1930 as heat transfer fluids, flame retardants, lubricants, and particularly as insulating fluids in electrical transformers. The technically desirable properties of PCBs, however, contribute to their environmental persistence and stability, and we know now that PCBs have been detected in almost every component of the global ecosystem, from the Antarctic to the Arctic. While there are no known diseases due to environmental exposure (fish in the Great Lakes may contain up to 50 ppm), about 1200 people were poisoned in Japan in 1968. Rice oil had been contaminated with PCBs from a heat exchanger, and people suffered from skin lesions and diseases of the nervous, reproductive, and endocrine systems. In experimental animals, PCBs have caused cancer of the liver, but no conclusive, direct relationship between cancer and human exposure has ever been made.

The evidence concerning the presence of PCBs in the environment, in the food chain, and in the human body, and the possibility of PCBs causing cancer led to the decision to restrict the use of PCBs as much as possible, and to develop control mechanisms to eliminate the release of PCBs into the environment.

The use of PCBs has been prohibited for several years, now, and during this time it has become clear that there was much more toxic material than just PCBs in the heat exchanger fluid that contaminated the food in the famous outbreak of symptoms in Japan. Because the fluid had been used a long time at high temperatures, it had toxic quaterphenyls and dioxins present, as well. Also, since the ban on PCBs, it has been found that several kinds of bacteria actually break down the PCBs fairly rapidly.

DIOXINS AND FURANS

Dioxins and furans are shorthand terms used to describe two large families of chemicals. The correct names are *polychlorinated dibenzo* dioxins and *polychlorinated dibenzo* furans; these many compounds have nothing to do with diox*ane*, nor with furans in general, and must not be confused with these very different chemicals. Again, the terms dioxins or furans, when used in the usual way today, refer to large families of highly chlorinated compounds, and as in other extended families, the individual members differ a great deal in their "friendliness." In fact, the least toxic members are only one one-thousandth as toxic as the most toxic members of the same family. Because most popular media refer to any and all dioxins with a standard phrase "deadly dioxins," it is important to find out just what member or members of the family are really being referred to in each case. The dioxins and the related PCBs and furans all show considerably different toxicity to different animals, and some toxicologists are of the opinion that human beings are among the more resistant species in response to these chemical classes. These families of chemicals are not industrial chemicals in the sense of those deliberately produced — they are found only as unwanted side products from the production of other materials involving chlorine and phenols.

ESTROGENS IN THE ENVIRONMENT

It is not clearly established whether environmental estrogens affect people. However, considerable evidence suggests that such chemical pollutants may have a role in the reported increasing frequency of infertility, abnormal sexual organs, and particular diseases of the reproductive tracts of both males and females in diverse wildlife species. The synthetic chemicals thought to be involved include DDE (from DDT), some other older chlorinated pesticides (e.g., kepone), some PCBs, and phenols with a simple side chain of nine carbon atoms.

PULP MILL EFFLUENTS

The degree of toxicity of effluents from pulp mills, the nature of the toxic chemicals involved, and the ways to eliminate the toxicity are all puzzling questions. There is no doubt that the liquid effluent from pulp mills is toxic to fish. A great deal of research has been done; at first it appeared that mills using chlorine to bleach the pulp produced toxic effluent, but that those using other processes did not. Because the ecosystems receiving effluents from pulp mills are so complex, some simpler ways of predicting harm were needed. Sensitive and representative species and biological processes were sought to serve as barometers for impacts upon ecosystem health as a whole. These were called bioindicators, one of which is an enzyme called ethoxyresorufin-*o*-deethylase (EROD) that increases in fish exposed to pulp mill effluent. The level of EROD correlates well with pulp mill effluent exposure, but the adequacy of this test as a bioindicator of toxic effect is questionable. For example, in mink that drank BKME (bleached kraft mill effluent), EROD was clearly increased, yet no illness was produced. It became clear, mainly from studies in fish, that it is not the chlorination that makes pulp mill effluent toxic, but something that all pulp mills have in common — probably some natural component of trees that reaches streams only if the trees are ground up and extracted (pulped). Chlorination can lead to novel compounds in effluent, but no longer appears to be the main culprit, at least for the health of fish.

MERCURY

The uncontrolled release of mercury from chemical plants into water, and the unintentional consumption of seed grain treated with mercury, have caused several tragedies. In 1971 to 1972, more than 6000 people had to be hospitalized in Iraq after eating homemade bread prepared from wheat treated with a mercury fungicide. A few years earlier, central nervous system diseases, as well as brain damage in unborn children, were observed in Japan. This became known as Minamata disease, and proved to be due to eating contaminated fish. The fish had concentrated the mercury from the waste water of a chemical plant. After this episode, it was found that many North American rivers and lakes had also been contaminated with mercury, and "fishing for fun, but not eating" signs have gone up on many water bodies. Today, it is recognized that mercury levels in some lakes may be high due to natural causes.

ECOTOXICOLOGY TO ASSESS RISKS OF POLLUTING THE ENVIRONMENT

Ecotoxicology is an extension of toxicology that examines adverse effects upon populations and communities of organisms and upon the health of ecological systems, instead of the individual. Ecotoxicology deals both with direct effects upon the health of organisms, and also with indirect effects, in which the effect is produced through degradation of the environment. Ultimately, the assessment of ecotoxicity is at the level of populations, communities, and ecosystems, including both their structures and their functioning. Frequently the most sensitive or most vulnerable species is focused upon; sometimes the commercially important species and, often, the key primary producers and consumers are the focus. It is difficult to relate single species results, either in the laboratory or in the field, to the community or ecosystem level effects. A battery or panel of tests or set of species for field monitoring is an improvement. The measurement of ecosystem-level health impacts is further complicated by compensating mechanisms in the system. For example, the extermination of a species or population by a toxicant may have no effect upon the ecosystem functions, if another species took its place effectively. Thus, apparent ecosystem health could obscure a loss of species diversity, a decrease in genetic diversity, and a reduction of future resilience, as well as an accumulation of nonlethal burdens of diseases and potentially harmful gene mutations. There are many ecological principles that must be brought into consideration when moving from single-species toxicological testing or monitoring, to assessment of toxicity at the ecosystems level; books on ecology and on population dynamics should be consulted, if further details are desired.

SHOULD CHLORINE BE BANNED?

There is no doubt that many chemical molecules containing chlorine have caused toxic problems; nor is there any doubt that chlorine itself, as a gas, is highly toxic (it was used as a chemical warfare agent in World War I, with devastating results). Equally well known is chlorine's great effectiveness in making drinking water free of bacteria, parasites, and viruses. Chlorine is used to make swimming pools safe, and in very many commercial products, where the chlorine atom is part of a molecule, in which state it has none of the properties of the parent toxic gas. When toxic effects from pulp mill effluent were at first attributed to products resulting from the use of chlorine in the bleaching process, a number of environmental groups called for a ban on chlorine. It must be understood that, taken literally, such a ban would be disastrous as well as fundamentally impossible. Chlorine is one of the more abundant elements making up our world. In its ionized form (chloride) it is also essential to life. Consequently, whatever is meant by "ban chlorine," cannot be the broad, literal meaning. To make sense, it would have to be translated into a slogan like "ban industrial uses of chlorine" or "ban chlorination of organic molecules." But even these proposals would deprive us of several antibiotics, anesthetics, plastics, and so on. The complex analyses (risk/benefit, cost/benefit, and risk/alternate risk) that are necessary cannot be detailed here, but the reader is urged to think of the real meaning and implications of the catchy phrase "ban chlorine," not only in developed countries, but in other parts of the world with more contagious disease and fewer options.

UPSETTING THE NATURAL ENVIRONMENT

Sometimes an industry can produce chemical effects without deliberately making any particular chemical. One example is the flooding of land that contains normally insoluble mercury minerals. Under the flooded conditions, microbial action can liberate the mercury and convert it into methylmercury (a form which then accumulates in the food chain), resulting in contaminated fish. Another example is the moving of rocks from below ground to the surface or placing them on a shoreline. Depending on the chemistry of the rock, it may change from a stable form to a weathered, soluble form that can then cause toxic effects. A third example is the liberation of large quantities of natural chemical constituents of trees into the effluent from pulp mills. Among these chemicals from trees are phytoestrogens — chemicals which can alter sexual development in wildlife and perhaps in people as well. However, it is not clear whether the effects are primarily bad or good because phytoestrogens seem to reduce the risk of certain cancers and may counteract harmful effects of environmental estrogen substances from nonforest sources.

NEW CHEMICALS FOR INDUSTRY

Some examples of industrial research rapidly developing new chemicals are in the fields of microelectronics, superconductivity, and new solvents for water-based paints. The compounds developed include gallium arsenide; complex compounds of five or six different metals including, for example, strontium; butyrolactones; and various lactams. These novel chemicals have been tested for their toxicity, of course. Yet history suggests that chemicals with extraordinarily favorable properties for industrial uses may in the long run prove to have unexpected negative effects upon human or environmental health. These chemicals have been known for some time in the laboratory, but their role as industrial chemicals is new. Consequently, they cannot have a "track record" of total safety. Recent research shows that chemicals made up of several different metal atoms have a toxicity different from the sum of the toxicity of each metal alone. In addition, a new type of toxicity may be emerging in which a chemical able to dissolve oils in water and nontoxic in traditional tests nevertheless helps other, unrelated chemicals, to penetrate through the skin. As another example (this time in the enthusiastic pursuit of safety for human health), some alternative pesticides have been developed which are extremely safe for people, but highly toxic to minute pond animals essential as food for fish and ducks. On the other hand, equally vigorous pursuit of environmentally friendly chemicals (e.g., odorless, water-soluble wood finishes) may result in the introduction of novel toxic hazards for people's health. A certain vigilance is prudent even after new chemicals enter commerce. A balance is needed between healthfulness for humans and healthfulness for the environment.

SAFE USE OF INDUSTRIAL CHEMICALS

In our thirst for more and more goods and conveniences, we have overlooked learning how to handle chemicals properly, from production and use to disposal. Recently, initiatives like Workplace Hazardous Materials Information System (WHMIS) in Canada, and its counterpart in the U.S. (the 1982 Generic Standard issued by the Occupational Safety and Health Administration; OSHA), have seriously tackled this gap. Such programs, nationally mandated, provide for classification of chemicals according to various kinds of potential hazard; standardized hazard symbols and labeling; compulsory provision of basic information on each chemical produced, shipped or used; plus mandatory training programs in chemical safety. There is growing recognition that there are no safe chemicals, only safe ways of manufacturing, handling, and using them. Safety measures should, therefore, form an integral part of any operation involving chemicals. Exposure prevention and control should constantly have the highest attention in all handling of chemicals, the more so when there are doubts about potential long-term adverse effects. If we are bound to err, let's err on the side of caution.

Recently, the manager of a big chemical manufacturing plant in Switzerland named ignorance, misunderstanding, and information gaps as the most important sources of hazard to his employees. This statement should be taken seriously, not only by the chemical industry, but also by public authorities as well as all users of chemicals, whether farmers, craftsmen, home-makers, or do-it-yourselfers.

Chapter **11**

WASTE CHEMICALS

WHAT ARE WASTE CHEMICALS?

Wastes are chemicals of a kind that are not wanted, in a particular place, at a particular time. Wastes are, nevertheless, chemical resources: sources of needed materials, or of fuel energy. Every process for producing something we need, such as food, clothing, houses, equipment, and drugs, also creates waste. Waste chemicals are substances that are discarded because the producer (generator) or owner has no further use for them. These chemicals may be by-products of the production process, or they may be materials that have served their purpose. Although many of these materials are harmless, some may be hazardous or dangerous.

Hazardous waste chemicals are wastes that, by their nature and quantity, are potentially harmful to plants, animals, or humans and to the environment. These wastes can be solids, liquids, sludges (thick mixtures of liquids and solids), or gases. They may be poisonous, flammable, explosive, or corrosive. Given their pronounced physical differences, such hazardous chemicals all require special disposal techniques to eliminate or reduce the hazard they pose. A waste is hazardous primarily because of its identity and the quantity of chemicals it contains. Unfortunately, wastes generally contain more than one chemical. These may interact, thus producing new hazardous materials. In addition, some chemicals are more toxic in the presence of others, e.g., mercury toxicity is increased with trace amounts of copper.

PROBLEMS WITH HAZARDOUS WASTE

Millions of tonnes of hazardous waste chemicals are generated each year worldwide. Hazardous wastes are being recognized as one of the major environmental and societal problems of the late 20th century.

The proper disposal of hazardous wastes is a serious problem which has resulted from decades of inadequate waste disposal. Society as a whole must share the responsibility for this because the driving force behind the very existence of the waste problem is thirst for consumer products and eagerness to purchase them at the lowest short-term market cost. This philosophy frequently has resulted in generating large quantities of dangerous by-products. Society has become highly dependent upon chemicals. Technology is progressing more quickly than society's ability to deal with the wastes produced. Certainly in the short run, as we operate automobiles, paint houses, grow food, make and dye cloth, or print books, substantial amounts of hazardous wastes will still be generated. Technology has developed more quickly than our understanding of it and faster than our social control mechanisms, thus permitting society to become far more dependent on chemicals than most of us realize.

WAYS OF DEALING WITH WASTE CHEMICALS

The ideal solution would be to eliminate or reduce the amount of waste generated. Massive reduction in waste generation is most feasible for newly constructed plants. In some areas, for example, "zero effluent" pulp mills are being proposed. However, only a limited number of changes can economically be made to older facilities in the pursuit of this goal.

Modern management of hazardous waste includes five major but overlapping categories: pretreatment, deep well injection, incineration, landfill or burial, and resource recovery.

Hazardous wastes can be, and should be, pretreated or detoxified to improve the efficiency, economics, and safety of subsequent disposal. In some pretreatment technologies, wastes are so completely detoxified they may not need further treatment or monitoring. Such methods include biological, chemical, or physical pretreatment.

Deep well injection is another method used for disposal of wastes. It involves pumping or draining liquid wastes through injection tubes into highly porous rock formations, at depths to several hundred meters. There is no doubt that this is not a permanent disposal option, because contamination of the surrounding environment may occur.

Incineration is probably the safest and most effective method of disposal for most types of hazardous wastes, except those with high concentrations of noncombustible and heavy metals, such as lead, mercury, or cadmium. Complete incineration, however, is essential and involves complex, sophisticated, and costly technology, expensive antipollution devices, and close ongoing monitoring of performance.

Burial in secure landfills is still seen as an option for disposal of waste chemicals. Landfills are claimed to be "secure" when designed to prevent contamination of surface and ground water and constructed to be impervious to external sources of water and to prevent accidental leakage of toxic washout. The question remains, however, whether any landfill, however ideally constructed and monitored, can ever be guaranteed truly secure, on a very long-term basis. Opponents of this method suggest that, sooner or later, toxic leachate (washout) may escape and contaminate surrounding soil or water sources.

The recovery and reuse of commercially valuable materials from hazardous industrial wastes is the preferred disposal method, short of not generating any waste. Many substances can be recovered and then reused by the producer of the wastes for another process, or piped or transported to another industry which may need this material. This is one of what we might call the "five Rs" of modern responsible waste management: **rethink** (the whole system), **reduce** (waste at its origin), **refine** (the process for greater efficiency of materials and energy use), **reuse** (effluent, containers, etc.), and **recycle** (to other uses). One industry's waste is another industry's raw material. For example, scrap from a meat packing plant is rendered down and reacted with amines to produce a flotation agent for refining minerals. Closer to home, community paint exchanges keep down costs at the same time they keep much solvent out of landfills, back lanes, and sewers.

ALWAYS HAZARDOUS OR ALWAYS SAFE: CAN WASTES CHANGE?

Wastes need to be looked at in two ways. What kinds of atoms are present? And what types of molecules are present? No matter what is done to the waste (biological treatment, high temperature incineration, etc.), exactly the same number of the same kinds of atoms will come out of the treatment as went into it! Even burning at high temperatures does not destroy the toxicity, if the toxicity is based on toxic atoms. On the other hand, the molecules (i.e., spatially linked groups of atoms) after treatment may be changed quite drastically, and are usually no longer toxic.

The atoms of the chemical elements fall into two groups: those that are biologically hazardous in and of themselves (radioactive isotopes and toxic elements); and those that are not. Some molecules are particularly toxic because they contain atoms of toxic elements. However, the majority of toxic chemicals are made up of atoms that are practically nontoxic; the toxicity of such chemicals comes from the exact way in which the atoms are linked together and arranged in the molecule. This situation is like a door key: if you melt it, the same amount of the same metal will still be there, but it will not be able to unlock the door. When assessing hazards from wastes and looking at ways of managing or treating them, it is very helpful to keep in mind three questions: does this waste contain atoms that are themselves toxic? Where will the toxic atoms go? What will happen to the toxicity that depends on molecular structure?

WASTES THAT CHANGE FROM HAZARDOUS TO SAFE

When toxicity is due to the presence of atoms that are themselves toxic (e.g., mercury or cadmium), permanent loss of toxicity is not possible. Nevertheless, temporary decrease or blocking of such toxicity can occur. One example is chloride (which is abundant in seawater) making mercury or cadmium ions much less available to living things — and thereby less toxic in marine than in freshwater environments. Other examples are chemicals exuded into the soil from plant roots, which also decrease metal toxicity, temporarily.

When toxicity is due instead to the molecular arrangement of atoms that are themselves practically nontoxic, then their toxicity may be reduced or destroyed in many ways: by biochemical reactions in bacteria and other living things; by sunlight; by reactions on soil particles; and by burning at high temperatures.

WASTES THAT CHANGE FROM SAFE TO HAZARDOUS

Waste rock that is highly insoluble and generally inert (such as some sulfide ores) can turn toxic. If such rock is brought to the surface or dumped along a shore, it can become oxidized and thereby soluble, newly available to living organisms and toxic to them. Another example is household garbage, which will increase greatly in toxicity if it is burned in a smoky or smouldering fire. In these fires there are many heat-promoted chemical reactions and conversion of simple carbon compounds into more complex ones, such as the polycyclic aromatic hydrocarbons (PAHs), many of which are real cancer hazards. This kind of reaction, in wood and coal fires, results in the accumulation of PAHs in chimneys. In earlier days, chimney sweeps actually went inside chimneys to clean them, and got covered in soot. It was the great excess of scrotal cancers among chimney sweeps that led to the discovery of chemical carcinogenesis.

WHERE ARE THE HAZARDOUS WASTES?

Hazardous wastes can be found almost anywhere: from the upper atmosphere to deep beneath the ground, and from highly industrialized cities to utter "pristine" wilderness. The geography and mapping of wastes is very important, from the local scale (e.g., around abandoned gas stations) to world scale (e.g., for the greenhouse gases or the ozone holes). Hazardous wastes can be found in the atmosphere, in water bodies, and on or below the land surface. Some wastes are very localized; others move all over the planet.

In order to estimate the risks to human health and to the ecosystem, it is necessary to know the distribution of the hazardous materials, their probable routes of travel, and where the people or other species of concern are located. Are the chemicals coming from a pipe or chimney ("point source"), or from a diffuse operation like fertilized fields ("nonpoint source")? How will features of the landscape influence the direction and speed of movement of the material? And where are the centers of population, or critically sensitive species located? These geographic factors, together with the toxic potential of the waste chemicals themselves, determine risk, and consequently mapping is an important activity in managing toxic wastes.

Hazardous wastes may be roughly classified as to where they are found: in the air, in water, or on land.

HAZARDOUS WASTES IN THE ATMOSPHERE

Important aspects of atmospheric hazardous wastes include acid rain, the "Greenhouse Effect" and Global Warming, ground-level ozone excess, and ozone holes in the upper atmosphere.

Acid Rain

Acid rain is a waste chemical problem. Smoke stacks of coal-fired power plants release not only ash, but also sulfur dioxide and nitrogen oxides. The sulfur dioxide emissions have become known as "acid rain." Acid rain has the ability to damage the environment and many living and nonliving things.

Since the Industrial Revolution began, humanity has contributed more and more substances to the atmosphere. Frequently included are large amounts of sulfur and nitrogen oxides. These oxides may occur naturally in the environment, but are mainly a result of human activity, such as burning fossil fuels for heat or power, smelting sulfide ores, or producing exhaust from automobiles. These oxides tend to be changed by complex processes in the atmosphere to sulfuric and nitric acids, which then fall to the ground as acid precipitation in rain, snow, or even dry particles.

The Greenhouse Effect and Global Warming

Certain gases accumulate in the upper atmosphere, reflecting heat (infrared) rays back to Earth, instead of letting them pass freely into space. This "Greenhouse Effect" at a certain natural level is essential to maintain the world's climate, and to lessen the day/night temperature swings. Since the Industrial Revolution, however, the number and amount of such gases have risen ever more rapidly. The projected result is a progressive rise in temperature, brought about by carbon dioxide, methane, chlorofluorocarbons (CFCs), and nitrous oxide — the main "greenhouse gases." The steady increase in the efficiency of the "gaseous greenhouse" is expected to produce global warming, which appears (despite some controversy) to be causing more concern in the 1990s than is acid rain.

A very considerable fraction of the total greenhouse gas emission still comes from nature, not from industries. Indeed, most of the methane is from living systems, and the carbon dioxide comes from metabolism of carbon compounds as well as from the exhaust of internal combustion engines and open burning.

Ozone

Ozone is a trace gas which, at ground level, is harmful, but in the upper atmosphere, is beneficial or even essential to health on Earth. Sunshine directly from outer space is extremely rich in high-energy ultraviolet rays (UVA, UVB, and UVC), which damage nucleic acids and proteins, producing mutations, rapid aging, and blindness. Normally the ozone layer acts as a natural filter, keeping sunlight's most damaging UV radiation from penetrating to the surface of our planet. However, when chlorinated fluorocarbons (CFCs) with molecules light enough to reach the upper atmosphere react with oxygen, chlorine atoms are released, and the naturally occurring ozone layer at that altitude is depleted. This reduction in the amount of ozone is called an "ozone hole," and occurs chiefly over the antarctic and arctic regions. If ozone holes persist over populated areas, then skin and eye troubles, mutations, and cancers can be expected.

Ozone is a "plus," then, in the upper atmosphere; quite the opposite is true at ground level. At the surface of the Earth, nitrogen oxides (which also contribute to acid rain) are very prominent among the chemicals involved in the production of ground-level ozone. The ozone, together with the nitrogen oxides themselves, is highly irritating to the respiratory tract and accounts for numerous cases of suffering in humans.

Although ozone smells "fresh" and was even used years ago as a room air freshener, it is highly reactive chemically, giving rise to "free radicals" that damage many essential components of living cells. Some natural compounds and many synthetic ones, in the presence of solar UV and atmospheric oxygen, produce photochemical smog rich in ozone and "supercharged" forms of oxygen, which are quite toxic to people, animals, and plants. Curiously, a *deficiency* ("hole") in the upper atmospheric ozone will probably *increase* the ground-level ozone, by letting more powerful UV reach the surface.

HAZARDOUS WASTES IN THE WATER AND "TOXIC BLOBS"

Some wastes are typically found in water, or are a special problem if found there. These chemicals either float on top of water, dissolve in it, are suspended in it, or separate out into liquid blobs or solid sediments on the bottom. The water bodies involved can be surface ones (ponds, creeks, lakes, rivers, oceans) or underground (water-table, aquifers, wells). One river, in Ohio, became so polluted that its surface caught fire, and it was declared a fire hazard! Water in other forms (rain, snow, fog, dew) can deposit, move, or release chemicals within the ecosystem. For example, dew releases some herbicides from a field's surface; and polluted fog can be especially toxic to vegetation on mountainsides. In colder climates, the water may form ice or snow covers over both water and land, further decreasing the rate of loss of hazardous compounds, by blocking both their evaporation and their destruction by sunlight.

"Toxic blobs" are an unusual form of contamination of water bodies. The best known, perhaps, are the ones that were found in the St. Claire River in Ontario, a highly industrialized waterway. These blobs consisted mostly of perchloroethylene, but since this solvent is not soluble in water, it acted like an extraction solvent, and appeared to remove other insoluble pollutants out of the passing river water, concentrating these other pollutants in the blob!

HAZARDOUS WASTES ON LAND

Most people are interested in two situations involving hazardous wastes on land: abandoned gasoline storage tanks (e.g., former gas stations with corroded underground tanks) and hazardous waste sites or chemical "dumps." There is a third situation where interest is developing: toxicants unexpectedly piggy-backing on useful and rather harmless materials.

Abandoned Gasoline Storage Tanks

These may not be the most serious hazardous waste sites, but they make up for that by being highly visible and very numerous. The usual problem with such sites is accumulation of various hydrocarbons in the ground around corroded and leaky tanks. There can be a fire and explosion hazard, of course, but the toxic hazards are typically less severe, unless there are significant amounts of benzene, or unless there has been entry into an aquifer. Usually, the soil and water are tested for the presence of "BTEX" (short for benzene, toluene, ethylbenzene, and xylenes), and any significant levels lead to more extensive testing for BTEX distribution and the possible presence of PAHs and other prioritized chemicals. The problem of abandoned gasoline storage tanks is part of two much bigger topics: abandoned hazardous wastes sites of many types; and the toxic hazard assessment of real estate properties for sale.

Hazardous Waste Sites and "Dumps"

In the past, the most common disposal methods for wastes was to pour them into landfill sites or "nuisance grounds" or pits. Many of these pits did not have a lining to prevent seepage of these substances into the soil. Frequently, there was little or no treatment to reduce these dangers. Once unsafely discarded, many of these substances do not remain immobile. They have the ability to move through the environment, contaminating soil, water, and air.

Examples of this may be seen, in the U.S., in the "Valley of the Drums" in Tennessee and Love Canal, at Niagara Falls, New York. Thus, a major problem facing society today is to ensure that dumped wastes do not cause health and environmental damage. It is an enormous task to rectify the mistakes of the past. Ways for safe disposal of waste chemicals must be found. Mistakes must be prevented in the future, even if cleanup and prevention may be very costly for everyone.

A U.S. Environmental Protection Agency (EPA) survey of hazardous waste disposal sites has tabulated nine major categories of hazards or impacts from such sites:

- Ground water contamination
- Forced closures of wells
- Habitat destruction
- Human health problems
- Soil contamination
- Fish kills
- Livestock losses
- Inoperable sewage systems
- Other effects (damage to crops or wildlife, air pollution, fire or explosion hazards of abandoned sites).

Toxicants Piggy-Backing on Otherwise Nonhazardous Wastes

Waste material, including mixed sewage sludge, is frequently used to improve the quality of poorer soils. This is a useful procedure. However, care needs to be taken to ensure that major sources of toxic elements do not enter the sewage that will contribute to sludge intended for use in this way. While the concentration of toxic metals or other toxic elements may not be high, toxic elements are not degradable, and can accumulate from year to year, with each successive addition of sludge. Similar "piggy-backing" of toxic elements can occur with materials which are not wastes at all: some natural sources of phosphate rock used in fertilizers contain appreciable amounts of cadmium, which can accumulate as these fertilizers are used year after year on the same land; and some natural sources of dietary calcium supplements (e.g., from some dolomite and oyster shell sources) may contain enough lead to pose a health risk, with heavy and long-term use. Still other examples of piggy-backing toxicants include transport of toxic chemicals on fine particles in water or air; unsuspected toxic by-products accompanying nontoxic products (e.g., the dioxins that once came from the synthesis of some herbicides); and the assisted penetration of some toxicants into skin, by otherwise highly useful special-purpose solvents (e.g., in some paint products).

CONTAMINATED SITES: ASSESSING THE RISKS AND REMEDYING THE SITUATION

Wastes are often complex mixtures, which can enter the environment by different routes and move in various directions at certain speeds. To assess the toxic risk in relation to a particular source, it is necessary to estimate five things:

- The potential hazard of the chemicals present
- The pathways they could take to the nearest, most susceptible, or valuable organisms
- The probable doses to be received by those organisms
- The timing, frequency, and duration of exposure, and
- The susceptibility of the organisms (e.g., infants).

The risk can be calculated from all of the factors just listed, but usually it is a complex function. There are several approaches to assessing toxic risk: one is based upon chemical measurement of a large number of chemicals on a particular list. Another is the "Toxics Identification and Evaluation" (TIE) approach, in which tests for effects on living systems (biological assays or bioassays) are used to detect toxicity in samples, and only then are chemical analyses done to identify and measure the chemicals responsible. Chemical analysis alone, for everything on an official list, will be costly and still may miss a significant toxicant present, but not yet on an official list. Addition of bioassays for toxicity makes the chemical assessment more focused, and makes it unlikely that a significant but unlisted chemical will escape notice.

Once the risk assessment has been completed, it should be clear whether or not the situation needs remediation. If, for toxic outcomes, the added exposure risk does not exceed one in 1,000,000, over a lifetime, then most risk estimators would consider the risk too low for remediation to be necessary. If the risk is considered sufficiently high, then remediation could take the form of either removal of contaminated soil or water for processing at another location, treatment on site, or a combination of these. The treatment, depending upon the chemicals and the circumstances, could be:

- Spreading and composting ("land-farming wastes")
- Pumping through a chemical reactor
- Pumping through a biological purification (biodegradation) system
- Treatment right within the contamination, by injecting bacteria able to destroy the offending compounds (*in situ* bioremediation).

Waste chemicals tend to migrate from their source, carried in air, surface water, ground water, or on eroding surface particles. The patterns of their movements are called "plumes." There are a number of systems for immobilizing (containing) a migrating plume of contamination; these can be used in conjunction with the treatment methods just described.

It is important to recognize that there can be unexpected and unidentified chemicals in a particular waste or location; that natural products converted to unnatural forms or moved from their natural locations can be a toxic hazard; and that the disappearance of any identified toxic compound does not always mean that the toxicity has gone — conversion to another form has at times provided an even more toxic product.

To reduce the likelihood and severity of future mistakes, and to prevent contamination of the environment with the possible resulting health hazards, the minimization, recovery, and reuse of waste chemicals should become standard practice.

PREVENTING FUTURE MISTAKES

The need for more careful management of waste chemicals and for international cooperation to achieve this goal is evident from the many reports about the rather large areas contaminated by toxic materials. Both transportation and disposal of hazardous wastes present major problems for society. To alleviate these problems, careful surveillance must be undertaken from the time these materials are produced until they are discarded or recycled. What is needed is a Cradle-to-Grave approach (described in Chapter 10: "Industrial Chemicals"). This approach seeks to manage chemicals from the time of their research and development (cradle) to their proper disposal (grave), or their recycling.

No country should ever be the dumping ground for another country's waste. International cooperation in monitoring transborder movement of toxic chemicals is essential and more joint efforts toward finding safe disposal methods are necessary. Knowing where and in what volumes hazardous wastes are located in a given region would be advantageous for firefighters, emergency measures personnel, and the environment department spill response teams. The common goal is to ensure the safe and efficient management of waste chemicals. We cannot afford to repeat past mistakes.

Chapter 12

INDOOR AIR

Widespread construction of energy-efficient (or energy-conserving) buildings started in the 1970s. Such building construction reduces heating and cooling fuel requirements and costs and, incidentally, reduces the extent of migration of outdoor pollutants into the interior of buildings. Nevertheless, there is a negative side to this environmentally friendly mode of construction.

Making buildings energy efficient in some cases decreases the quality of indoor air. Increasing insulation, adding weather-stripping, etc., can reduce energy consumption, but without adequate air exchange, the air within a building becomes stagnant and will retain any pollutants generated within it. This has resulted in a new health problem, termed "Sick Building Syndrome" or "20th Century Disease." A person experiencing this problem is unable to live or work in an energy-conserving building without developing symptoms. Sometimes, as in people with "Multiple Chemical Sensitivity," reactions to polluted indoor air can be life-threatening. While the number of people having these extreme reactions is considered to be small, there are a variety of less severe health effects which can be linked to indoor air pollutants. Some symptoms that have been reported include irritation of the respiratory tract, eyes, throat, and skin; rash; nausea; diarrhea; fatigue; difficulty concentrating; headaches; difficulty breathing; lightheadedness; dizziness; and fainting spells. Some indoor air pollutants include combustion gases and particles, formaldehyde, radon, and asbestos. A general overview is given in Table 12.1.

Multiple Chemical Sensitivity, also known as "total allergy syndrome" or "environmental illness," usually occurs following an acute environmental exposure to a chemical, sometimes a solvent or pesticide. After initial exposure, an individual can become sensitive to very low level chemical exposures and experience symptoms in more than one part of the body.

The existence of Multiple Chemical Sensitivity remains highly controversial, primarily because of the lack of an established mechanism to explain how exposure to concentrations of a chemical that are well tolerated by the population at large can produce the array of symptoms seen in "sensitized" individuals. Frequently symptoms vary among sensitive individuals, and different organ systems are affected.

It has been suggested that the mechanism causing Multiple Chemical Sensitivity can be a disorder in the regulation of neurogenic inflammation.

Inflammation is an abnormal condition of redness, swelling, heat, and pain localized in a specific tissue of the body. Neurogenic inflammation is inflammation caused through an action on the nervous system.

COMBUSTION GASES

Carbon monoxide and nitrogen dioxide are the two major combustion gases of concern in indoor air quality. Carbon monoxide is a highly toxic, colorless, odorless, tasteless, nonirritating gas, which is a by-product of the burning of fossil fuels. Nitrogen dioxide is a highly toxic, irritating gas. Indoor sources of these gases are unvented gas ovens, stoves, and furnaces, gas water heaters, gas or kerosene space heaters and wood or coal-burning stoves and fireplaces, and cigarette smoking. Improperly maintained natural gas furnaces and collapsed chimneys are also sources of carbon monoxide, as is vehicle exhaust from a garage attached to a building.

TABLE 12.1
Indoor Air Pollutants

Pollutant	Sources	Possible Health Effects

Combustion gases
  Carbon monoxide
  Nitrogen dioxide
	Kerosene heaters; wood stoves; unvented gas stoves; attached garages	Headache; dizziness; nausea at low concentrations; at higher concentrations neurological effects; lung damage and disease; can be fatal

Formaldehyde	Urea formaldehyde foam insulation; plywood; particle board; furniture; curtains; carpet	Nose, throat, and eye irritation; nasal cancer in experimental animals
Radon	Earth and rock beneath home; well and spring water	Responsible for 5 to 20% of all lung cancers
Asbestos	Some wall, ceiling, and pipe insulation; heat shields (asbestos paper)	Skin irritation; cancer and lung disease, especially after extensive exposure
Combustion particles	Tobacco smoke; wood smoke; unvented gas appliances; kerosene heaters	Nose, throat, and eye irritation; respiratory infections; heart and respiratory disease; emphysema and lung cancer

Both nitrogen dioxide and carbon monoxide, though in different ways, interfere with the body's ability to supply oxygen to the tissues. Nitrogen dioxide is less of an acute hazard since high (potentially lethal) levels usually are not reached in a home. Very high concentrations of carbon monoxide (greater than 1000 ppm) are quickly fatal. Carbon monoxide is responsible for many deaths each year. If the victim survives, neurological changes can persist for weeks or even years. Exposure to carbon monoxide can result in a variety of adverse health effects (Table 12.1), and is dependent on the concentration of the gas in the body. Certain populations, such as pregnant women, infants, and people with anemia or cardiovascular and respiratory diseases, are more susceptible to the effects of carbon monoxide. However, all individuals should limit their exposure to the gas. In fact, there should be less than 1 ppm of carbon monoxide in buildings. At levels higher than 1 ppm, the source of the gas should be identified and the problem corrected.

Combustion gases can be significantly reduced (approximately 70%) in the home by ensuring that a sufficiently large fresh air intake is available (preferably with an air-to-air heat exchanger for energy conservation purposes). Additionally, all gas appliances should be properly adjusted, well ventilated, and checked for leaks and particle buildup. This also applies to fireplaces and wood-burning stoves. Nor should vehicles be allowed to idle for extended periods of time inside an attached garage.

FORMALDEHYDE

Formaldehyde is a pungent, colorless, water-soluble gas. It is primarily an indoor-generated pollutant with a variety of sources including plywood, particle boards, urea formaldehyde foam insulation, adhesives, varnishes and lacquers, wallpaper, carpets, curtains, and furniture. Mobile homes tend to have higher formaldehyde levels than conventional homes since they are usually constructed with more plywood and particle board. Lesser sources are gas and wood stoves and tobacco smoke.

Most people can detect formaldehyde, by smell, at levels of 1 ppm or less in the air. Levels of 2 to 3 ppm generally cause mild irritation of the eyes, nose, and throat. Few people can work or live comfortably under such conditions for long. Levels of about 10 ppm cause the eyes to water; levels of 10 to 20 ppm cause breathing difficulties and coughing. Long-term exposure to formaldehyde may result in structural and functional changes in the lungs. In addition, long-term exposure to high concentrations of formaldehyde has been shown to cause nasal cancer in animals.

Indoor air formaldehyde levels may be reduced by increasing the air exchange rates and using a dehumidifier. Low-formaldehyde particle board and exterior grade plywood (which releases less formaldehyde than interior grade) are commercially available and may be useful in building construction and renovation.

RADON AND RADON DECAY PRODUCTS

Radon is an odorless, colorless, radioactive gas which is produced as a result of the spontaneous breakdown (so called "decay") of radium. Radium is a naturally occurring element which is found in small concentrations in the soil and rock everywhere in the Earth's crust. Concentrations vary widely, depending upon geographical location. Soil under a home is the principal contributor to indoor radon levels. Radon gas can enter buildings in the following ways: from the soil through cracks and openings in the foundation, walls, and floors; through untreated spring or well water; and from earth-derived masonry material such as concrete.

Radon decay products are capable of attaching to respirable particles and may become embedded in the lungs where they irradiate the surrounding tissue. The greatest risk from long-term radon exposure is lung cancer. Based on the incidence of lung cancer among uranium miners (who are exposed to very high radon levels), it is believed that radon causes 5 to 20% of all lung cancers.

Control of radon entering a building through the basement can best be achieved by sealing cracks in the foundation and between the foundation, walls, and floor. In addition, radon concentration can be reduced by increasing the ventilation in the basement and crawl space.

ASBESTOS

Asbestos is a naturally occurring mineral fiber formerly used in construction materials. Asbestos was present in insulation for walls, ceilings, pipes, and boilers. Asbestos-containing paper, paper tape, floor and ceiling tiles, textiles, gloves, protective gear for firefighters, do-it-yourself brake repair materials, and asbestos cement board were also commercially available. These products were used to protect the floors and walls around wood-burning stoves as well as to insulate hot water pipes and heat ducts. The most commonly mined and commercially used types of asbestos are the serpentine class (e.g., chrysotile) and the amphibole class (e.g., crocidolite, anthophyolite, amosite, actinolite, and tremolite).

In humans, the serpentine chrysotile asbestos is less harmful than the amphibole asbestos. This is likely because chrysotile breaks down more readily and completely in the body than do the amphiboles. The longer asbestos persists in the body, the more damage it causes. Amphibole fiber concentrations increase with duration of exposure, but chrysotile concentrations do not.

Asbestos can cause three forms of lung disease in humans: asbestosis, lung cancer, and malignant mesothelioma. Asbestosis results from the formation of fiber-containing (fibrous) tissue in the walls of the alveoli (thin-walled chambers of the lungs). Asbestosis is characterized by shortness of breath, cough, tightness in the chest, and pain. Malignant mesothelioma is a type of cancer in which there is formation of tumors of the cells covering the lung surface (mesothelium). This is a very rare disease and is primarily associated with asbestos exposure. Both mesothelioma and asbestosis usually do not occur until at least 20 to 30 years have passed since exposure.

Fiber size (i.e., length and diameter) determines the potential for toxic effects associated with asbestos exposure. Shorter fibers are more completely removed from lung tissue than long ones. Fibers greater than 3 μm (microns) in diameter do not readily penetrate lung tissue. Asbestosis is caused by shorter fibers (2 μm or less in length); mesothelioma usually results from exposure to longer fibers (2 to 5 μm) that are less than 0.5 μm in diameter; and other types of cancer associated with the lung occur from exposure to asbestos fibers longer than 10 μm.

Left alone, intact asbestos poses little health risk. Asbestos exposure, however, can occur in buildings following the aging and cracking of asbestos-containing products during their physical breakdown and from cutting or breaking the material during renovations. It is recommended that asbestos-containing material be left alone if it is in good condition. It has been shown that removing the asbestos increases the asbestos fiber content of the air substantially. If the asbestos must be removed, it is best to hire a professional certified in asbestos removal. Very specific guidelines are available governing the proper removal of asbestos, safely. It has been suggested that simply coating the intact asbestos with a good quality paint may be preferable to removal.

Chapter **13**

SMOKE

The chapters in this book vary in length. This one is obviously very short. It stands alone because smoke is such an important topic. All fires produce smoke, which may be thick or thin, light or dark. Exposure to smoke occurs from, among other sources, forest fires, burning buildings, industrial fires, burning wastes or stubble fields, and even volcanoes.

Smoke is defined as a complex mixture of aerosols and gases produced when material undergoes thermal decomposition. A wide range of chemicals is produced in smoke resulting from burning materials, either natural or synthetic. As an example, burning Douglas fir wood (a natural material) produces more than 75 different chemicals in the resulting smoke. When materials burn, chemicals such as carbon dioxide, carbon monoxide, nitrogen oxides, sulfur dioxide, and ammonia, plus many others are produced. The types and amounts of thermal decomposition products formed are dependent upon temperature, rate of heating, oxygen availability, and whether or not there are flames involved. Smouldering fire and black smoke indicate the production of many new organic chemicals.

Contributing to the burden of "greenhouse gases" is just one of the many environmental hazards that come from burning various materials. Toxic metals or phenols and dioxins are driven into the air when preservative-treated wood is burned; and polycyclic aromatic hydrocarbons (PAHs) may be produced from burning stubble fields, as well as in exhaust from diesel and gasoline engines.

SMOKE FROM FIRES

Smoke has always been with us, often making people sick. Any appreciable quantity of smoke causes irritation, and reduces light and visibility.

Contrary to popular belief, most fatalities from fires are not due to burns, but to inhalation of toxic gases produced during combustion. An estimated 80% of fire deaths are due to inhalation of gases and smoke. Victims who survive often experience acute and delayed respiratory problems, headaches, dizziness, nausea, and excessive thirst.

There are two main types of chemicals in smoke: asphyxiants and irritants. Asphyxiants produce a deficiency of oxygen in the blood and an increase of

103

carbon dioxide in the blood and tissues (asphyxia). Irritants cause irritation to skin, eyes, and the respiratory tract.

Individuals with preexisting respiratory diseases (e.g., asthma, emphysema, bronchitis) and cardiovascular diseases (e.g., coronary heart disease, ischemic and nonischemic heart disease) are considered at greater risk when exposed to smoke. Children are particularly at risk if they have respiratory problems.

Firefighters are at increased risk of developing respiratory diseases and possibly cancer, because of exposure to smoke. Acute effects experienced by firefighters include headaches, coughing, nausea, dizziness, irritation of mucous membranes, and breathing problems. Studies of firefighters show the chronic effects of greatest concern are lung disease (asthma, bronchitis, emphysema); heart disease (coronary heart disease, ischemic and nonischemic heart disease); cancer of the lung, brain, and nervous system; and possibly leukemias and cancers of the genitourinary tract, colon, and rectum. However, the link between cancers in firefighters and their occupational exposure is controversial. Increased mortality from aortic aneurysm has also been reported in firefighters. Aortic aneurysm is a defect in the aorta (large blood vessel leading directly from the heart) which causes enlargement of the vessel which may eventually rupture. This link, too, is controversial.

TOBACCO SMOKE

Smoke from tobacco is a primary source of two major classes of combustion products: respirable suspended particulates and a category of chemicals called polycyclic aromatic hydrocarbons. Respirable suspended particulates are very small particles which become lodged in the respiratory tract. The majority (96%) of these particulates and gases (such as carbon monoxide) produced by cigarette smoking are present in "sidestream" smoke. Sidestream smoke is defined as the smoke arising from the lighted end of a cigarette and passing directly into a room. Some polycyclic aromatic hydrocarbons, produced during incomplete combustion, are well-known cancer-causing agents.

Smoking tobacco products is of concern because of carbon monoxide, nicotine, tar, and cancer-causing chemicals (such as benzo(a)pyrene) released during their use. The main effect of carbon monoxide on the body is to greatly decrease the ability of blood cells to carry oxygen. This results in less oxygen in the body tissues where it is essential for life. The effect of nicotine on the body is on the nervous system, some hormone levels (adrenalin and noradrenalin), and on heart rate and blood pressure (increasing both).

The effects of tobacco smoke upon the smokers themselves are well established and include emphysema, heart disease, and cancer of the lung, oral cavity, esophagus, bladder, and pancreas. Of greater concern (especially in a chapter on smoke, itself) are the effects of "passive smoking." Passive smoking is the inhalation of smoke that has been exhaled by the smoker, plus sidestream smoke. What are the effects of passive smoking? Primary effects of short-term exposure are coughing, headache, nausea, and irritation to the eyes, nose and throat. People with allergies or cardiovascular or respiratory diseases are particularly sensitive to the effects of respirable suspended particulates. Respiratory problems in children, especially in those under 2 ycars of age, are more frequently present when one or both parents smoke. There is an increased incidence of lung cancer in the nonsmoking partners of smokers.

The use of tobacco has been shown to increase the activity of drug-degrading enzymes in the body. This means that smokers frequently require as much as 50% more of some prescription drugs to get the same effect as nonsmokers.

Women who smoke during pregnancy have an increased likelihood of miscarriage, premature delivery, and below normal weight babies at full-term delivery. Some people suggest that these children are more prone to Sudden Infant Death Syndrome (SIDS; crib death). This is a term usually used to describe deaths in infants from 1 to 6 months old and attributable to no known cause.

Chapter 14

POWER LINES, VIDEO DISPLAY TERMINALS, AND SUNLIGHT

Not only chemical agents, but also physical agents are included in the science of toxicology. Certainly, physical agents can harm living things, and endanger human health. The physical agent to be considered here is radiation (differing in forms and sources) from power lines, video display terminals, (VDTs), and the sun. Other physical agents capable of having toxic effects are dust, sound, pressure, and vibration, although these agents are not discussed here.

There are two major types of radiation that can cause toxic effects. These are ionizing radiation (e.g., X-rays and gamma rays) and nonionizing radiation (e.g., ultraviolet, visible light, infrared, microwaves and radio waves, and extremely low frequency electromagnetic fields). In this chapter, discussion will focus only on two types of nonionizing radiation: the extremely low frequency radiation produced by power lines or VDTs and the ultraviolet radiation produced by sunlight, electric arcs, sunlamps, and sunbeds. Neither the use of UV radiation for medical purposes nor the use of natural or artificial radioactive substances will be discussed.

ELECTROMAGNETIC FIELDS (EMFs)

TYPES AND SOURCES

Electricity, produced in power-generating stations, is transmitted to our homes, work places, schools, and other locations by power lines ranging widely in voltage: high voltage transmission lines (110,000 to 735,000 volts) to distribution lines (110 to 110,000 volts), depending on location. The term "power lines" refers to both transmission and distribution lines.

Two types of fields are produced by electricity, the electric field and the magnetic field. These fields are commonly referred to, together, as electromagnetic fields (EMFs). Electric fields are generated by any electrically charged body; magnetic fields are generated only when a current flows. The two field types have an important difference: electric fields are easily blocked (by almost all electrically conducting materials, including buildings and human bodies); whereas magnetic fields easily pass through almost all materials.

The strongest electric fields that the public encounters (at least in Canada and the U.S.) are directly under high voltage transmission lines. Outside the right-of-way of high voltage lines, the electric field strength usually does not exceed 1000 volts per meter.

Unlike electric fields, the strongest magnetic fields are usually associated not with high voltage transmission lines, but with certain occupations in which the worker is closer to motors or other electrical equipment for long periods. The sources of magnetic fields in residential areas can be divided into four general categories: high voltage transmission lines; distribution lines; building wiring; and electrical appliances. In fact, the magnetic field inside a house close to a high voltage transmission line may be no greater than in a house located at a distance. The reason is that the strength of the magnetic field in the house may depend much more on the distribution line to the house and the wiring in the house than on the fields coming directly from the transmission line. In the home, the strongest magnetic fields are found closest to some electrical appliances (e.g., can openers, hair dryers, electric razors). Such appliances can generate magnetic fields greater than those from transmission lines, but the fields decrease rapidly as one moves away from the appliance. As well, exposure to appliances is usually brief and intermittent. However, exposure to electric blankets and heated waterbeds is more prolonged and closer to the body; exposure to electric shavers and hair dryers is also very close to the body.

Electromagnetic fields are produced not only by power lines and by electric motors in appliances, but also by video display terminals (VDTs) such as the monitors used with computers. The EMFs produced by VDTs vary in intensity, depending on their design. There are differences in EMFs produced even among units of the very same model.

TOXICITY OF ELECTROMAGNETIC FIELDS

Concern has been expressed about adverse health effects resulting from exposure to EMFs from power lines and VDTs. The reported risk of adverse effects during pregnancy (e.g., miscarriages, birth defects) and the development of childhood cancers (especially leukemia and brain tumors) has been of particular concern. Evidence to date is controversial. Although some studies have shown a weak association between childhood cancers and exposure to EMFs, others have not. In addition, most of the studies have not shown that exposure to EMFs from power lines or VDTs causes miscarriages or birth defects.

As a group, studies that examined the association between adult cancers and residential exposure to EMFs show no consistent pattern of increased risk. In adults, the elevated risk of myeloid leukemia is seen only in the highest exposure category. The most consistent findings, to date, are increased risks for leukemia and brain cancers among "electrical" workers. Of additional concern is the reported increased risk of male breast cancer among electrical workers. This is an exceedingly rare disease and even a modest increase in the disease is cause for concern. When all the information is taken together, the evidence is suggestive (but not yet conclusive) for a carcinogenic effect of exposure to EMFs, particularly in children and electrical workers. The association is strongest for workers in various "electrical" occupations and weakest for adults in the general population.

The present situation of not knowing exactly what properties of the EMFs should be measured (e.g., frequency, voltage, variability) contributes to the controversy. However, this is an active area of research at the present time.

ULTRAVIOLET RADIATION

TYPES AND SOURCES

Sunlight is considered the most important source of human exposure to ultraviolet radiation (UV rays), but sunlamps and sunbeds used in tanning booths or tanning salons are also important sources. The types of UV rays considered most generally harmful to humans are UVA (long wave ultraviolet) and UVB (middle wave ultraviolet) rays. Short wave ultraviolet (UVC) is not normally encountered by most people, except in certain occupations (e.g., some laboratory work and arc-welding) where it is potentially very dangerous, producing blindness, mutations, and other damage among unprotected onlookers. Originally it was thought that only UVB and UVC rays were harmful, but now it is known that UVA rays are, too. In fact, UVA rays have been shown to penetrate the skin more deeply and damage the deeper tissues. One component of tissue that is damaged by UVA rays is collagen, a protein which provides support to the skin. Accumulated damage to collagen is one reason why chronically UV-exposed skin ages and wrinkles prematurely.

TOXICITY OF ULTRAVIOLET RADIATION

Ultraviolet radiation from sunlight, sunlamps, and sunbeds has been associated with the development of three types of skin cancer (basal cell carcinoma, squamous cell carcinoma, and malignant melanoma). Most skin cancers occur late in life as a result of cumulative UV exposure from childhood onwards. Basal and squamous cell carcinomas develop in skin that has been exposed repeatedly to UV rays, whereas malignant melanoma may develop on exposed or nonexposed skin. Malignant melanoma is the most serious type of skin cancer; early treatment is advisable, as this disease can be fatal.

The best way to detect skin cancer early is to examine the skin often. Seek immediate medical attention if any abnormally dark or discolored patches or spots develop in the skin or if any mole bleeds, crusts over, or changes in color, size, or shape.

UV rays may also cause allergic reactions and chronic damage to the skin such as premature wrinkling, aging, and dryness. Some medications (e.g., birth control pills, some antibiotics, high blood pressure medication, and tranquillizers) may increase one's sensitivity to UV rays. Contact with certain plants can do the same — that is, cause photosensitivity reactions — and some individuals are abnormally sensitive to sunlight, e.g., those with lupus erythematosus.

Not only do UV rays damage the skin directly, they also can damage the immune system and the eyes. Damage to the immune system results in decreased ability of the body to fight disease, whereas damage to the eyes may eventually result in cataracts. Children are especially vulnerable to the sun because their skin is thinner, more sensitive, and therefore less protected against the penetration of UV rays. All sunburns should be avoided, especially in childhood. Suntans, even without any sunburns, cause skin damage. After all, we would not turn brown unless the sunlight (or tanning salon light) changed our skin chemistry. Therefore, a sunscreen that protects against both UVA and UVB rays and has an SPF (sun protection factor) of 15 or greater should be used. In the tropics and at high altitudes, sunscreens with even higher SPF values are needed.

Chapter 15

CHEMICALS MISUSED

Using more of anything than is required, or using any chemical product in any way for which it was not meant, is chemical misuse. It is easy to misuse chemicals. Most times, it is done unintentionally because the risks are not known, label instructions are not followed, or people are so used to the product that they do not bother to recheck the label directions to see if they have changed and if the product is being used correctly and safely. Such misuse can have unfortunate effects upon our health, someone else's health, or the environment. Some misuse is directly under our personal control; some of it is possibly within our "sphere of influence" (e.g., at home, school, work place, or in our closer environment).

A second type of chemical misuse is deliberate, often referred to as "chemical or substance abuse." Here, abusers knowingly inflict negative effects of a chemical, primarily upon themselves, and only secondarily upon others. Nevertheless, this secondary effect can be devastating. The deliberate misuse of chemicals ranges from suicide and murder by poisoning, all the way down to drinking large excesses of coffee. Quite a range!

FROM HOUSEHOLD CLEANERS
TO ALL SORTS OF THINGS

Almost any chemical or chemical product can be misused. For example, anything listed in Chapter 9 ("Paints, Solvents, Cleansing Agents, and All Kinds of Things") has been misused or abused at some time or another. The categories of chemicals that are misused go on and on: paints, solvents, paintbrush cleaners, paint removers, antifreeze, deicer, kerosene, gasoline, glue, and products in aerosol (spray) cans. These are all found around the home, garage, farm, or cottage. They are all reasonably safe when used for their intended purpose and if they are used properly, that is: according to directions, with sufficient ventilation, if stored in the original container out of reach of small children, and so on. The list of common household chemicals that are sometimes misused continues with: detergents, bleaches, other cleaners, disinfectants, deodorants, drain cleaners, polishes and waxes, mothballs, perfumes, and shampoos. These products, too, are safe if used in the way intended, according to label instructions, and provided they are not mixed (in the case of cleaners and especially drain cleaners), nor transferred to unlabeled containers, left where children could misuse them, or disposed of in an unsafe manner.

Some other types of chemical misuse are discussed below.

OVERUSE OF PRESCRIPTION DRUGS

Although prescription drugs must be prescribed (or ordered) for an individual by a physician, the potential for overuse exists. When physicians prescribe medication, they take into account your state of health, sex, weight, age, and other medications (both prescription and nonprescription or "over-the-counter" drugs) you may be taking.

Overuse of a prescription drug occurs when more than the recommended amount is taken or if the drug is taken more frequently than prescribed. Twice as much is not twice as good and may, in fact, be harmful. Sharing your prescription with others is another way in which these drugs are misused. Just because they help one person does not mean someone else can be helped as well. When a prescription is shared, it is being prescribed without the experience of a medical education.

Prescription drugs that tend to be overused include tranquillizers, sedatives, painkillers (analgesics), antidepressants, and sleeping pills. Most of these drugs are prescribed for short-term use and, unless authorized by a physician, should not be taken for extended periods of time. For more information about prescription drugs, please refer to the books by the Canadian Medical Association, Smith, and the U.S. Pharmacopeial Convention, in the section, "Further Suggested Reading."

COFFEE AND ALCOHOL

Coffee and alcohol use and misuse may result in physical or psychological dependence. Physical dependence means that the body becomes used to having a particular chemical and goes into a withdrawal reaction if its use is stopped abruptly. Psychological dependence exists when a substance becomes so important to a person's thoughts, emotions, and activities that it is extremely difficult to stop using it. This condition is marked by a compelling need or craving for the substance.

Caffeine is the ingredient in coffee, tea, and cola drinks that is responsible for the pleasant feeling we may experience after drinking such beverages. Some people, however, become depressed when they use caffeine-containing products.

Alcohol is commonly used for its psychological effects, relaxation or euphoria (feeling of well-being), or sometimes for its intoxicating effects. As a result there is a misconception that alcohol is a stimulant. In fact, it is a depressant, which acts on the brain, first reducing our inhibitions, then as more alcohol is consumed, depressing all brain functions.

The use of alcohol by pregnant women can cause Fetal Alcohol Syndrome in the developing fetus. It has been suggested that even small amounts of alcohol during pregnancy can affect the fetus. Body size of the offspring is decreased compared to normal infants, the brain is smaller, and there are varying degrees of mental retardation.

The effects of alcohol are more pronounced in children than in adults because, in the young, bodily systems and organs are still developing. The inability to concentrate and learn are just two of the effects of alcohol that result because alcohol interferes with the brain and nerves. Alcohol also affects other organs and systems: the gastrointestinal tract (stomach and intestine), the liver, heart, muscles, blood, hormone levels, mouth, throat, and lungs.

TOBACCO

The use of tobacco is hazardous to human health, whether smoked or smokeless products are used. More information on smoked tobacco is provided in Chapter 13, "Smoke."

The use of smokeless forms of tobacco (commonly referred to as "spit tobacco"), such as snuff and chewing tobacco, is growing rapidly, particularly in adolescent and young adult males. Chewing tobacco is usually sold as leaf tobacco (packaged in a pouch) or plug tobacco (in brick form), and both are placed between the cheek and the gum. Snuff is a powdered tobacco (usually sold in cans) that is placed between the lower lip and the gum. Both release nicotine quickly into the bloodstream.

Health risks associated with the use of spit tobacco include gum and tooth disease and bad breath; nicotine addiction; oral lesions, including leukoplakia (precancerous formation of white, leathery patches in the mouth); oral cancer (including the lip, tongue, cheek, and throat), which could lead to removal of parts of the lip, cheek, or face; and heart disease due to the nicotine which increases heart rate, blood pressure, and sometimes causes irregular heart beat, all of which may lead to greater risk of heart attacks and stroke. Thus, the use of smokeless tobacco is not a safe alternative to smoking.

STREET DRUGS

When street drugs are used, they cause chemical changes in the body, especially in the brain and nervous system. These changes may produce pleasant or unpleasant feelings. The reaction of a person to street drugs varies greatly from one individual to another, and from time to time in the same user. The effects depend on the amount used, individual personality, previous drug use, and on the physical and social environment at the time of use.

The idea that drugs obtained from natural sources are less harmful than synthetic ones is not correct. Chemicals are either dangerous or they are not — whether they grow in fields or are manufactured in the laboratory. Many plants are toxic and can cause serious consequences if used inappropriately. Another consideration is that street drugs often contain unknown ingredients, including glass, weed killers, and more potent drugs.

In addition to the information provided below, more information about street drugs can be found in the books by Hindmarsh and by Scott and Hindmarsh, identified in the "Further Suggested Reading" section of this book.

CANNABIS

Cannabis (marijuana, hashish, or hash oil) comes from the plant, *Cannabis sativa*. Cannabis, sometimes referred to as "pot," is the most frequently used street drug. The plant contains over 400 chemicals. Upon smoking, these 400 chemicals are changed into approximately 2000 different chemicals. Most of the research has been done on only one of these substances, the mind-altering (psychoactive) ingredient called THC (tetrahydrocannabinol). The amount of THC found in marijuana today is about 10 to 15 times greater than it was a few years ago. Furthermore, the THC from smoking one cigarette (joint) stays in the body for up to 28 days. Weekend users are in danger of build-up of THC in the body. Cannabis affects the synapses (junctions between cells, where impulses are transmitted) of the brain and slows down transmission of signals. This is one reason why cannabis impairs coordination, memory, and judgment. It interferes with learning and the ability to perform tasks, such as driving a car.

Cannabis smoke contains about 70% more of the cancer-causing chemical, benzo(*a*)pyrene, than does cigarette smoke. In addition, it contains about twice as much tar as the strongest cigarette. This results in damage to the lungs, decreasing their ability to use oxygen efficiently. Lung changes observed are similar to those seen in older people who have smoked tobacco for decades. Medical reports are now describing mouth and throat cancer in cannabis smokers who are only in their 20s.

Long-term effects of cannabis, in addition to precancerous changes and cancers of the respiratory tract of heavy users, include a weakened immune system, which means the body has a decreased ability to fight off diseases. Cannabis also impairs the hormonal and reproductive system. It decreases the levels of some hormones in the body, especially sex hormones in young males. In females, changes in the menstrual cycle can occur, and the use of cannabis during pregnancy can affect the development of the baby in the womb. Besides these early developmental effects, cannabis also has been found to alter the genetic code and chromosomes in the offspring of users.

COCAINE

Cocaine is a very powerful drug which stimulates certain activities of the brain and other parts of the nervous system. It comes from the leaves of the *Erythroxylon coca* bush.

The effects of cocaine are similar to "speed" (methamphetamine), but wear off much faster. Therefore, cocaine must be used more frequently to get the same effects. Cocaine users have experienced very unpleasant feelings of intense anxiety and panic, increased heart rate, high blood pressure, and elevated body temperature. Users who "snort" (sniff) cocaine frequently develop serious damage of the sensitive inner lining of the nose and the nasal septum. Higher doses of cocaine can cause mental illness, such as severe paranoia and hallucinations. Behavior becomes strange and violent. Death results from overdose.

Recently, it has been found that reverse tolerance develops from repeated cocaine use. Tolerance is the need for larger and larger amounts of cocaine to get the same effect. For some unknown reason, the body reverses this tolerance effect suddenly and the larger dose becomes a toxic dose, which can result in severe epileptic seizures and death.

In 1986, a new method for using cocaine made headlines. The cheap and deadly "crack" cocaine was produced. Smoked in glass pipes, rather than snorted (the most common method for using cocaine), crack provides immediate effects, lasting for 5 to 20 minutes. The assault on the body and brain occurs more swiftly and more profoundly than with cocaine. The immediate health effects are sore throat and hoarseness, and heavy use leads to emphysema of the lungs. An acute overdose can cause breathing to stop; heart rate and blood pressure increase, leading to risk of heart attack. The effects on the brain are the same as with ordinary cocaine use. The appetite is suppressed, leading to weight loss and malnutrition. It is difficult for the nonaddict to imagine the depth and viciousness of depression that an advanced cocaine addict suffers from, and accidents and suicide are often the result.

CRYSTAL METH ("ICE")

In the late 1980s, the drug commonly known as "ice" or "crystal meth" (crystals of methamphetamine) became available. Crystal meth is a cunning drug, maneuvering itself from a quick smoke with an energized, long-lasting high to something akin to demonic possession. These crystals are the amphetamine equivalent of crack cocaine. The effects of inhaling ice are almost as instantaneous as those from crack. The "high" appears to last much longer than with crack. The higher and longer the high, the more severe the depression that follows once the effect of the drug wears off.

SOLVENT ABUSE

There is a fine line between inhaling enough solvent to obtain a "high" and inhaling too much and passing into unconsciousness. Sudden deaths are associated with inhalation of volatile solvents, present in some common household products. This is because such solvents can sensitize the heart to a normal body chemical, adrenalin. Emotional and/or physical stress can cause the body to release extra amounts of adrenalin as well. This extra adrenalin makes the heart beat so rapidly that the body cannot cope, and frequently death results, unless medical intervention is available.

Another dangerous abuse habit is inhaling propane and butane. These gases freeze the mouth and throat causing edema (fluid accumulation) and may result in breathing difficulties and death.

HEROIN AND OTHER OPIATES

Heroin is one of the most powerful members of a class of drugs called narcotic analgesics (painkillers). Other members of this class are codeine and morphine, which occur in nature and are obtained from the opium poppy. Heroin is obtained by chemical modification of morphine. All opiates can cause addiction and when discontinued result in severe withdrawal symptoms. Because the purity and potency of street heroin is rarely known, overdose is frequent and often results in death.

LYSERGIC ACID DIETHYLAMIDE (LSD)

Lysergic acid diethylamide (LSD; "acid") is one of the most powerful hallucinogenic drugs known. Doses as small as 40 to 100 micrograms per person can produce a number of physical and psychological effects. These amounts are so small, they are barely visible to the naked eye and would fit on the head of a pin. It takes only microgram amounts of LSD for a "hit." A microgram is one millionth of a gram.

The effects of LSD and similar hallucinogenic drugs (such as mescaline, "angel dust," psilocybin or "magic mushrooms") include changes in emotions, thinking, memory, behavior, and the way things are seen (misperceptions and hallucinations). In addition, LSD produces nausea, anxiety, and muscle tremors. The effects of the drug wear off after 4 to 12 hours.

A phenomenon called "flashback" may occur after LSD use. Flashbacks, which can be triggered by stress, are recurring states of altered consciousness and perception, and may include severe hallucinations occurring long after drug use is discontinued. Injuries and deaths have resulted from incidents (e.g., feeling invincible and jumping off buildings or trying to stop traffic) that occur during flashbacks.

LOOK-ALIKE DRUGS AND DESIGNER DRUGS

Two newer sets of drugs are now on the street. These are "look-alike" drugs and "designer" drugs. Look-alike drugs, as the name suggests, look the same as other available drugs, but may not actually contain that drug at all. In other words, "you do not get what you think you got (and paid for)." Designer drugs (e.g., "China white," "Persian porcelain") are prepared by "underground chemists" who discovered that by slightly changing the chemical structure of an existing drug, a new, temporarily legal, drug can be created. Hence, the name "designer" drugs.

Designer drugs are essentially unknowns, and the human user is actually a guinea pig on whom they are tried first. Frequently, these drugs are several thousand times more potent than the drug from which they were designed. Consequently, the odds of taking an overdose are great, and death is frequently the result. An additional concern is irreversible movement disorders (like Parkinson's disease) produced in some users.

STEROIDS

Some athletes, competitive and noncompetitive, use anabolic steroids to improve their performances, a practice referred to as "doping." Some young people begin using steroids in junior high school. A wide range of serious health problems can occur. Aggressiveness and bouts of rage; cholesterol accumulation leading to heart attack and the necessity for heart bypass surgery; liver damage; and kidney failure are some that have been reported.

DELIBERATE USE OF CHEMICALS AGAINST PEOPLE: CHEMICAL WEAPONS

Chemical weapons turn civilization on its head: suffering and disease are not fought, but carefully cultivated. Scientists in chemical weapons laboratories use their knowledge of the functions of the body to devise even more effective means of halting those functions. Modern nerve gases were originally designed to kill lice, mosquitoes, or other insect pests, to help mankind. Now, in the hands of some countries, all these chemicals are, literally, "pesticides for use against people." Chemical warfare is, as one writer put it, "public health in reverse."

A BRIEF LOOK AT THE "WITCHES' RECIPE BOOK"

There are many chemicals that are used as chemical weapons. Some are clearly designed to kill ("lethal agents"); others are meant to harass or incapacitate people, or to destroy the plants that support daily life.

Lethal Agents

Lethal agents can be classified as causing one of the following effects: choking, blistering, destruction of blood function, damaging of nerves, other diseases, or death.

Choking agents are phosgene and chloropicrin. Phosgene damages the lungs, causing death by drowning in the body's fluids; chloropicrin acts in much the same way, but also causes vomiting, colic, and diarrhea.

Blister agents are so-called "nitrogen mustard" or "mustard gases," lewisite, and some similar compounds. Besides damaging the lungs, they cause skin rashes and blisters and destroy the bone marrow.

Blood function-destroying agents are hydrogen cyanide, cyanogen chloride, and cyanogen bromide. These chemicals block oxygen use in all parts of the body, causing rapid death.

Nerve agents, such as tabun, sarin, soman, and "VX," are direct relatives of the organophosphorus insecticides discussed in Chapter 4. They can cause death in 1 to 10 minutes. Only incredibly small quantities are needed to cause death.

Toxins are natural agents causing disease or death. They are derived from toxic plants, animals, bacteria, and fungi. Some examples are the very potent botulinus toxins (from the bacterium, *Clostridium botulinum*), staphylococcus toxins (from the bacterium, *Staphylococcus aureus*), ricin (derived from the castor bean plant), saxitoxin (obtained from shellfish), and trichothecene mycotoxins. The trichothecenes are produced by various fungi and were used against the Hmong in Laos (1975 to 1985) and also in Cambodia and Afghanistan. They do not cause death immediately, but death occurs after 24 hours.

Incapacitating Agents

"BZ" and LSD (lysergic acid diethylamide) are highly potent drugs which cause behavioral changes, accompanied by blurred vision, altered perceptions, hallucinations, fainting, and vomiting.

Harassing Agents (for instance, "tear gases")

"CN" and "CS" cause the eyes and the nose to water, cause a burning sensation in the throat, make it difficult to breathe, and can burn moist skin — but they do not kill. Others, like adamsite (or its relatives) cause sneezing, coughing, headache, shortness of breath, nausea, and muscular weakness.

Anti-Plant Agents (herbicides)

The herbicides 2,4-D; 2,4,5-T; picloram; and cacodylic acid are used as anti-plant agents. Mixtures of 2,4-D and 2,4,5-T became known as "Agent Orange" during the Vietnam War. The reasons for using these herbicides were: (a) to destroy the protective canopy of forested areas to make the enemy more visible and vulnerable, and (b) to kill plants used for food. Similarly, bromacil can be used in chemical warfare as a soil sterilant, preventing the use of gardens and fields to raise crops, for some considerable time.

Biological Agents

Viruses that produce a number of less commonly occurring diseases are said to have been used as warfare agents. These include the arthropod-borne viruses (Yellow Fever, tick-borne encephalitis, Japanese encephalitis, Dengue, Venezuelan equine encephalitis, Chikungunya, Rift Valley fever), and other viral infections not borne by arthropods (influenza, smallpox).

TOWARDS INTERNATIONAL CONTROL

The story of chemical warfare demonstrates that discoveries made in the cause of human welfare can be used to devise even more potent instruments of death. Our increasing ability to understand the delicate mechanisms which make life possible contributes also, unfortunately, to the knowledge of how to kill "better" and "more efficiently."

The 1925 Geneva Protocol, which banned the use but not the possession of chemical weapons, and the 1972 Biological Weapons Convention, which bans even the possession of biological weapons, including toxin weapons, are in place. But, repeatedly since 1925, history has shown that several countries have not hesitated to use such weapons if they thought using them would provide a military advantage.

The Conference on Disarmament concluded negotiations on a Chemical Weapons Convention which was opened for signature in 1993. This is the first international agreement that bans the development, production, stockpiling, and use of a whole category of "weapons of mass destruction." Under this agreement, not only are all chemical weapons and all such production facilities being destroyed under supervision, but all government and industry activities that could be turned to chemical weapons production are now liable to monitoring and inspection.

GLOSSARY

absorption: the movement of a chemical into the bloodstream or living tissue after it reaches the body.

acetaminophen: one of the nonprescription analgesic drugs used to relieve pain and fever (e.g., Tylenol®).

acetates: usually refers to solvents of the ester class, made by reacting acetic acid with an alcohol.

acetone: a volatile solvent.

acetylcholine: a substance produced in many nerve cells, which functions as a chemical messenger (neurotransmitter) to signal another nerve-, gland-, or muscle cell.

acetylcholinesterase: an enzyme found in mammals and many other organisms. It inactivates acetylcholine (a neurotransmitter).

acetylsalicylic acid (ASA; Aspirin®): one of the nonprescription drugs used to relieve pain and fever and to reduce inflammation.

acute: lasting a short time (e.g., 14 days).

aerosol: a suspension of very small particles of a liquid or a solid in a gas.

aflatoxin: a mycotoxin (see below) produced by *Aspergillus flavus* (a fungus) in peanuts and other commodities. Aflatoxin is a known carcinogen.

Agent Orange: a mixture of the phenoxy herbicides 2,4-D and 2,4,5-T used as a defoliant in the Vietnam War.

alcohol: usually refers to grain alcohol, i.e., ethanol, but there are many other kinds of alcohols, all of them more toxic than ethanol.

algicide: a pesticide used to kill or prevent growth of algae.

aliphatic: in chemistry, not aromatic; having no benzene or similar "aromatic" rings in the molecule or part of a molecule in question.

allergy: exaggerated or unusual reaction (such as sneezing, respiratory distress, itching, or skin rashes) to substances that do not produce comparable effects in most people.

Ames test: a test performed on bacteria to determine if a chemical can cause mutations.

analgesic: a drug used to suppress or relieve pain.

anemia: a lack of red blood cells.

anesthetic: a chemical which can block out pain and other sensations (e.g., of heat, cold, touch).

anionic: capable of releasing acidic ions.

antagonism: an interaction between two chemicals in which one chemical lessens the effect of the other.

antibiotic: a drug used to inhibit or kill microorganisms.

antihistamine: a type of drug used to decrease the amount of mucus secretion and provide some relief from a "runny nose."

aquatic: from or relating to water.

aqueous: watery; pertaining to a water solution.

aromatic: (technical meaning) compounds that contain one or more benzene rings.

arsenic: a naturally occurring toxic element.

aspirate: to inhale liquid, such as vomited material, into the lungs.

atom: the smallest unit of a single element that still maintains the physical and chemical characteristics of the element.

Bacillus thuringiensis (B.t.): a bacterium used as a biological insecticide, e.g., Thuricide®*.

benign: harmless.

benzene: the simplest of the aromatic hydrocarbon chemicals.

bioaccumulation: concentrating of a chemical by a biological system.

biochemical: a chemical produced by a living organism.

biodegradable: capable of being broken down to simpler chemicals by a biologic process or organism.

borates: salts or other forms derived from boric acid (also called "boracic acid").

borax: a form of sodium borate.

butyrolactone: a specialty solvent only recently introduced widely; an ingredient in some wood finishes and other products.

cadmium: a naturally occurring toxic element.

caffeine: a naturally occurring stimulant found in coffee, tea, or cola nuts.

cancer: a malignant growth of potentially unlimited size, invading local tissues, and often spreading to more distant areas of the body.

carbamate: a class of synthetic pesticides, sharing a particular cluster of atoms (e.g., carbofuran, carbaryl).

carbon tetrachloride: an organic solvent, heavier than water and insoluble in it.

carcinogen: a substance capable of causing cancer (see cancer).

carcinogenic: capable of causing cancer (see cancer).

cardiovascular: pertaining to the heart and blood vessels.

cationic: capable of releasing alkaline ions.

caustic: able to corrode skin and many other materials.

chemical: any specific substance composed of chemical elements such as oxygen, nitrogen, hydrogen, etc.

---

* Registered trademark of Sandoz Agro Canada, Inc. In Canada: Mississauga, ON.

chlorinated: having one or more atoms of chlorine present in a compound.

chlorine: an abundant chemical element.

chromosome: one of the group of structures that form in the nucleus of a cell during cell division. Chromosomes, bearing the DNA, carry the genetic code for the organism.

chronic: of long duration.

-cide: a suffix meaning "killer."

colic: acute abdominal pain.

coma: state of unconsciousness from which an individual cannot be aroused.

compound: a chemical substance composed of molecules all of the same kind.

congenital: pertaining to a condition existing before or at birth.

convulsions: abnormal and involuntary shaking of the body.

corrosive: able to physically damage, erode, or destroy the surface with which it has contact.

cresols: certain members of the phenol family of chemicals.

DDT: dichlorodiphenyltrichloroethane, a chlorinated hydrocarbon pesticide, no longer permitted for general use in most developed countries.

dermatitis: skin inflammation.

detergent: any synthetic cleanser that acts like soap.

dichlorobenzene, *para*: a solid chlorinated aromatic hydrocarbon used in mothballs.

2,4-dichlorophenoxyacetic acid (2,4-D): a phenoxy herbicide which selectively kills broad-leaved plants.

diethylstilbestrol: a synthetic drug acting like a sex hormone, formerly used as a growth-promoter in cattle and as a drug during pregnancy in humans, but since discovered to be carcinogenic.

dioxins: a family of chemicals, the most toxic of which is TCDD (see below).

DNA: the biochemical (deoxyribonucleic acid) from which genes are made; the chemical basis of heredity.

dosage: the size, frequency, and number of doses.

dose: the quantity of a chemical administered at one time.

dose–response curve: a graph of the relationship between the dose administered and the effect produced.

ecology: the branch of science concerned with the interrelationship of organisms and their environments.

ecosystem: a complex consisting of a community and its environment functioning as an ecological unit in nature.

edema: swelling of tissues (like skin) due to collection of fluid in them.

element: a chemical substance composed of atoms all of the same kind.

embryo: an organism in the early stages of its development. In the human, it is from conception to the end of the second month of uterine life.

embryotoxic: toxic to the embryo.

emetic: an agent that induces vomiting.

enzyme: a biochemical (protein) that speeds up the rate of a particular biochemical reaction.

epidemiology: the science that studies the cause, distribution, and control of epidemics or other diseases in a region.

essential oil: a concentrated oily extract characteristic of its plant source. It is not "essential" in the sense of an essential vitamin, etc.

ethyl alcohol (ethanol): the alcohol of alcoholic beverages.

ethylene glycol: a chemical used as antifreeze in motor vehicles.

excrete: to eliminate or get rid of by removal to the exterior.

fetotoxic: toxic to the fetus.

fetus: the later stages of a developing mammal in the womb. In the human, this refers to the period of uterine life from the end of the second month until birth.

food chain: a sequence of organisms in which each feeds on the member below it. The bottom member is a plant or alga.

fungicide: an agent that kills fungi, e.g., molds, mildews, or mushrooms.

gas: the highly mobile state of a substance that has a boiling point below normal room temperature.

gastric: pertaining to the stomach.

gastrointestinal: pertaining to the stomach and intestines.

gene: the smallest subunit of a chromosome or DNA chain that contains a genetic message.

genotoxic: damaging to genetic material of living organisms.

germicide: an agent that kills germs (microorganisms that cause disease).

habitat: the place in which an organism naturally lives.

half-life: the length of time required for the quantity of a compound or property in question to be reduced by one-half.

hazard: anything that presents a danger. Toxicants are hazardous because they have the potential to cause illness or death.

herbicide: an agent that kills plants.

hormone: a biochemical which is secreted by one organ in the body, and exerts an influence on a biochemical function or organ(s) somewhere else in the body, having been transported there by the blood stream.

hydrocarbon: an organic compound containing atoms of only carbon and hydrogen.

hydrocephalus: an abnormal increase in the amount of fluid within the cavity inside the brain.

hydrogen sulfide: a gas produced from manure by bacteria, generally smelling strongly of rotten eggs, but at high concentrations able to numb the sense of smell and thus seem odorless. Hydrogen sulfide is also a constituent of "sour gas" and sewer gas.

ibuprofen: one of the nonprescription analgesic drugs used to relieve pain, fever, and inflammation (e.g., Advil®, Motrin®).

inorganic: containing no carbon atoms.

insecticide: an agent that kills insects.

Integrated Pest Management: a systematic program of managing pests by using combinations of methods: natural enemies (parasites, predators, and pathogens — the three "p's"); culture practices; resistant varieties; and pesticides. The aim is to ensure both economic and ecological damage are minimized.

in vitro: literally "in glass" (test tube) and, more generally, simply outside a living organism (e.g., studies done in the laboratory without using living animals).

in vivo: in a living organism.

irritation: a purely local or topical reaction which may include redness, blistering, swelling, burning, or itching.

isopropyl alcohol (isopropanol): one of the alcohols used as a disinfectant.

isotopes: forms of the same element, but differing in atomic weight; they may be either stable or radioactive.

kerosene: one of the products obtained by distilling crude petroleum; it is a fuel and solvent.

lethal: causing death.

lipids: certain organic chemicals, insoluble in water and commonly known as fats and oils.

lipid-soluble: capable of being dissolved easily in fat or in solvents that dissolve fat.

lupus erythematosus (systemic): a disease of the connective tissue, with disordered function of the immune system.

malformation: a defective or abnormal structure or form.

malignant: very injurious or deadly.

mercury: a naturally occurring toxic element.

metabolism: the sum total of the biochemical reactions that a chemical undergoes in an organism.

metabolize: to produce biochemical change by the action of an organism and its enzymes.

methyl alcohol (methanol; wood alcohol): sometimes a component of antifreeze, shellac, varnish, and paint remover.

methylene chloride (dichloromethane): one of the simplest of chlorinated hydrocarbon solvents.

molecule: the smallest unit of a compound that still retains the physical and chemical properties of the compound.

morbidity: the frequency of illness in a population.

mortality: the frequency of death in a population.

mutagen: a substance capable of producing genetic change (e.g., mutations).

mutagenic: able to produce genetic change.

mutation: a genetic change.

mycotoxin: a toxic substance produced by a fungus.

naphthalene: a specific solid aromatic hydrocarbon with a characteristic pungent odor, used in mothballs.

nasal decongestant: a class of drugs used to relieve a stuffy nose and other symptoms of colds and allergies.

nausea: stomach upset accompanied by a feeling that one is about to vomit.

neurotransmitter: a product of a nerve, used to signal the next cell what to do (a chemical messenger).

nitrogen dioxide: a gas which is a deep lung irritant capable of producing pulmonary edema if inhaled in sufficient concentrations.

N-methyl pyrrolidone: a specialty solvent only recently introduced widely as an ingredient in many products.

non-Hodgkin's lymphoma (NHL): a particular kind of cancer of the lymphatic tissue.

nonionic: not able to release ions — neither acidic ones, nor alkaline.

nontarget: any organism not intended to be killed or damaged; usually in reference to the use of pesticides.

organic: broadly defined as anything containing carbon. Now frequently used to describe crops raised without pesticides or other synthetic chemicals.

organochlorine: among pesticides, belonging to the chlorinated hydrocarbon type (e.g., DDT, aldrin, dieldrin).

organophosphorus: belonging to a class of synthetic pesticides that contain phosphorous and that inhibit acetylcholinesterase (e.g., malathion; diazinon).

over-the-counter (OTC) medicine: those drugs available from a pharmacy without a prescription (e.g., ASA, acetaminophen, and many others).

oxalic acid: one of the naturally occurring organic acids.

ozone: the gas O_3, the most reactive form of oxygen.

***para*-dichlorobenzene:** see dichlorobenzene, *para*.

paraquat: one of the herbicides in the bipyridyl family of chemicals.

perchloroethylene (tetrachloroethylene): a chlorinated hydrocarbon solvent used in drycleaning and degreasing.

pesticide: an agent used to kill pests. The category "pesticide" contains many -cides; among others, this group includes all the insecticides and herbicides.

petrochemical: a chemical derived from petroleum.

petroleum distillates: substances obtained from the distillation of petroleum (e.g., kerosene).

pharmacokinetic: pertaining to the rates of absorption, distribution, metabolism, and excretion of drugs in the body.

phenols: a family of chemicals containing hydroxylated benzene rings; the hydroxy group makes these chemicals acidic and antiseptic.

pine oil: a complex mixture of oily material extracted from the wood of certain pines.

piscicide: an agent used to kill fish.

placenta: the organ that allows nutrients, oxygen, carbon dioxide, and wastes to be exchanged between the mother and her developing offspring, before birth.

poison: a chemical that is very highly toxic acutely. In law, a chemical with an oral LD_{50} of 50 mg/kg or less.

polychlorinated biphenyls (PCBs): a common term to describe a large family of chlorinated aromatic hydrocarbon chemicals formerly used as electrically insulating fluids in capacitors and transformers.

pyrethroids: synthetic compounds related to pyrethrins and resembling them in insecticidal properties (e.g., deltamethrin).

qualitative: pertaining to kind or type.

quantitative: pertaining to amount or degree — i.e., measured.

radionuclides: certain forms (radioactive isotopes) of specific chemical elements which emit radiation through atomic processes of various kinds.

reproductive toxicity: an effect that alters the normal reproductive function of an organism (e.g., loss of fertility).

resins: natural or synthetic gums or similar polymeric materials, including plastics.

risk: probability (likelihood) that harm will occur under specific circumstances.

risk assessment: process which evaluates the probability that harm will occur under specified conditions.

risk management: the process of dealing with (eliminating or minimizing, monitoring, and communicating) risks identified by risk assessment.

risk perception: probability of harm, as it appears to the individual or group concerned.

rodenticide: a chemical or product that kills rodents.

sediment: accumulated materials at the bottom of water or other liquid.

sensitization: the initial exposure of an organism to a specific antigen (protein, chemical) affecting the immune system in such a way that subsequent exposure induces an allergic reaction.

sewer gas: a mixture of gases, predominantly hydrogen sulfide, produced by sewage.

silo-fillers' disease: a disease in humans associated with the release from silage of nitrogen dioxide gas which, when inhaled, produces irreversible lung damage.

sodium hypochlorite: a form of chlorine bleach.

solanine: a toxin that occurs naturally in green skin on potatoes due to exposure to sunlight; also in other related plants.

sour gas: a kind of natural gas rich in sulfides.

sulfides: compounds of sulfur, with just one other element.

surfactant: an agent that reduces surface tension; see also detergent.

syndrome: a recognized combination of symptoms that occur together.

synergism: an interaction between two chemicals in which one chemical enhances the effect of the other.

synthetic: made by people; commonly used to mean not occurring naturally.

target organism: the organism intended to be killed or disabled by a pesticide.

TCDD: see 2,3,7,8-tetrachlorodibenzodioxin.

teratogen: a toxicant capable of producing birth defects.

teratogenic: able to produce birth defects.

2,3,7,8-tetrachlorodibenzodioxin (TCDD): the most toxic member of the family of dioxins.

thalidomide: a drug that was used as an antinauseant and sleeping aid in the 1960s and was subsequently shown to cause malformed limbs (phocomelia).

therapeutic: used in the art of healing.

toluene (methyl benzene): a common organic solvent.

toxicant: a toxic agent; a poison; see also toxin.

toxicity: the adverse biological effects produced by a chemical or physical agent.

toxicokinetic: pertaining to the rates of absorption, distribution, metabolism, and excretion of toxicants in the body.

toxicology: the study of the adverse effects of chemical and physical agents on living systems; the basic science of poisons.

toxin: a poison produced by a living organism; the word is very often used, interchangeably, to mean the same as toxicant. Strictly speaking, this is not correct.

trace metals: any metals present in very small amounts; some kinds are required for maintaining good health, e.g., zinc and cobalt.

trichloroethane: one of the chlorinated hydrocarbon solvents.

2,4,5-T (trichlorophenoxyacetic acid): one of the phenoxy herbicides.

turpentine: a solvent distilled from pine resin.

volatile: readily converted to the form of a gas or vapor.

xenobiotic: foreign to life.

xylene: chemical family name for dimethyl benzene, an aromatic hydrocarbon solvent.

FURTHER SUGGESTED READING

Ashby, E., *Reconciling Man with the Environment*, Stanford University Press, Stanford, 1978.

Berube, B., *Understanding Canadian Nonprescription Drugs. A Consumer's Guide to Safe Use*, Key Porter Books, Toronto, 1993.

(The) Canadian Green Consumer Guide, prepared by the Pollution Probe Foundation in consultation with Trayner, W. and Moss, G., McClelland and Stewart, Toronto, 1989.

Canadian Medical Association, *Guide to Prescription and Over-the-Counter Drugs*, The Reader's Digest Association (Canada), Montreal, 1990.

Carson, R., *Silent Spring*, Fawcett, Greenwich, CT, 1962.

Cockerham, L. G. and Shane, B. S., *Basic Environmental Toxicology*, CRC Press, Boca Raton, FL, 1994.

Dadd, D. L., *The Nontoxic Home*, Jeremy P. Tarcher, Los Angeles, distributed by St. Martin's Press, New York, 1986.

Derelanko, M. J. and Hollinger, M. A., Eds., *CRC Handbook of Toxicology*, CRC Press, Boca Raton, FL, 1995.

Ebbert, S. and Wilson, D., Part I: Household products, in *The Natural Formula Book for Home and Yard*, Wallace, D., Ed., Rodale Press, Emmaus, PA, 1982, 4–102.

Ecobichon, D. J., *The Basis of Toxicity Testing*, CRC Press, Boca Raton, FL, 1992.

Ehrlich, P. R., Ehrlich, A. H., and Holdren, J. P., *Ecoscience. Population, Resources, Environment*, Freeman, San Francisco, 1977.

Fairbairn, G. L., *Will the Bounty End?*, Western Producer Prairie Books, Saskatoon, 1984.

Fairbairn, G. L., *Canada Choice. Economic, Health and Moral Issues in Food from Animals*, Agriculture Institute of Canada, Ottawa, 1989.

Freedman, B., *Environmental Ecology. The Impacts of Pollution and Other Stresses on Ecosystem Structure and Function*, Academic Press, San Diego, 1989.

Government of Canada, *The State of Canada's Environment*, Supply and Services Canada, Ottawa, 1991.

Hamon, N. W. and Blackburn, J. L., *Herbal Products. A Factual Appraisal for the Health Care Professional*, Cantext Publications, Winnipeg, 1985.

Hamon, N. W., Blackburn, J. L., and DeGooyer, C. A., *Poisonous Plants in the Home and Garden*, Microtext Enterprises, Winnipeg, 1986.

Hart, J., Holdren, C., Schneider, R., and Shirley, C., *Toxics A to Z. A Guide to Everyday Pollution Hazards*, University of California Press, Berkley, 1991.

Hayes, A. W., Ed., *Principles and Methods of Toxicology*, 3rd ed., Raven Press, New York, 1994.

Health Canada, *Canada's Food Guide to Healthy Eating*, Cat. #H39-252/1992E, Supply and Services Canada, Ottawa, 1992.

Health and Welfare Canada, *Using the Food Guide*, Cat. #H39-253/1992 E, Supply and Services Canada, Ottawa, 1992.

Hindmarsh, K. W., *Drugs. What Your Kid Should Know*, College of Pharmacy, University of Saskatchewan, Saskatoon, 1989.

Humphreys, D. J., Poisonous plants, in *Veterinary Toxicology*, 3rd ed., Baillière Tindall, London, 1988, 209-281.

Hunter, L. M., Hazardous household products, in *The Healthy Home*, Rodale Press, Emmaus, PA, 1989, chap. 6, 104-130.

Institute of Medicine (Division of Health Promotion and Disease Prevention, Committee on Human Health Risk Assessment of Using Subtherapeutic Antibiotics in Animal Feeds), *Human Health Risks with the Subtherapeutic Use of Penicillin or Tetracyclines in Animal Feed*, National Academy Press, Washington, 1989.

Kamrin, M. A., *Toxicology. A Primer on Toxicology Principles and Applications*, Lewis, Boca Raton, FL, 1988.

Kamrin, M. A., Katz, D. J., and Walter, M. L., *Reporting on Risk. A Journalist's Handbook on Environmental Risk Assessment*, 2nd ed., Foundation for American Communications, Los Angeles, 1995.

Kendall, R. J., *Toxic Substances in the Environment*, 2nd ed., Kendall/Hunt, Dubuque, IA, 1983.

Kingsbury, J. M., *Poisonous Plants of the United States and Canada*, Prentice-Hall, Englewood Cliffs, NJ, 1964.

Klaassen, C. D., Ed., *Casarett and Doull's Toxicology. The Basic Science of Poisons*, 5th ed., McGraw-Hill, New York, 1996.

Kneen, B., *From Land to Mouth. Understanding the Food System*, NC Press, Toronto, 1989.

Mulligan, G. A. and Munro, D. B., *Poisonous Plants of Canada*, Agriculture Canada, Ottawa, 1990.

Naar, J., *Design for a Livable Planet. How You Can Help Clean Up the Environment*, Harper and Row, New York, 1990.

Ottoboni, M. A., *The Dose Makes the Poison. A Plain Language Guide to Toxicology*, 2nd ed., Van Nostrand Reinhold, New York, 1991.

Ray, D. L. and Guzzo, L., *Trashing the Planet. How Science Can Help Us Deal with Acid Rain, Depletion of the Ozone, and Nuclear Waste (Among Other Things)*, Regnery Gateway, Washington, 1990.

Rogers, S. A., *Tired or Toxic? A Blueprint for Health*, Prestige, Syracuse, 1990.

Schrecker, T. F., *Political Economy of Environmental Hazards*, Protection of Life Series Study Paper, Law Reform Commission of Canada, Supply and Services Canada, Ottawa, 1984.

Scott, S. and Hindmarsh, K. W., *Too Cool for Drugs*, Human Resource Development Press, Amherst, MA, 1993.

Smith, D. L., *Understanding Canadian Prescription Drugs. A Consumer's Guide to Correct Use*, Key Porter Books, Toronto, 1992.

Timbrell, J. A., *Introduction to Toxicology*, 2nd ed., Taylor & Francis, London, 1995.

U.S. Pharmacopeial Convention, Inc., *USP Drug Information, II: Advice for the Patient*, U.S. Pharmacopeial Convention Inc., Rockville MD, 1995.

Van Strum, C., *A Bitter Fog. Herbicides and Human Rights*, Sierra Club Books, San Francisco, 1983.

World Commission on Environment and Development, *Our Common Future*, Oxford University Press, Oxford, 1987.

INDEX

A

Acceptable Daily Intake (ADI)
20–22
Acetaminophen 52
Acetone 70, 76
Acetylcholine 32
Acetylsalicylic acid 5, 52
Acid rain 90, 91
Acids 53, 73, 75
Acute toxicity 5–7, 9, 10
ADI (see Acceptable Daily Intake)
Aflatoxin 17
Age, toxicity and 7, 22, 109, 112
Agent Orange 29, 118
Aging, and UV rays 91, 109
Air, toxicants in 16–18, 20–22, 35,
94, 100, 101, 103; indoor
97–101
Air Unit Risk 20, 21
Alar® 42, 47
Alcohol
ethyl alcohol 71, 72, 75, 112,
113
interactions with other
chemicals 30–31, 51, 52
reproductive and developmental
effects 11, 12, 112, 113
isopropyl alcohol (see also,
rubbing alcohol compound)
72, 74
methyl alcohol 70, 71, 74
rubbing alcohol compound 51, 53

Algae 9, 25, 45
Algicides 25
Alkali(s) 53, 72, 73
Alkaloids 60, 61, 63, 65, 66
Allergies, allergic reactions 4, 48, 49,
51, 52, 75, 97, 105, 109
Alternative
pest control methods 25, 84
testing methods in toxicology 9,
10
Ames test 13
Ammonia
anhydrous 41
from farming operations 44,
45
household 73, 74
in smoke 103
Analgesics 52
Analysis for toxicants 14, 19, 20,
94
Analytical toxicology 14
Antacids 53
Antagonism 8
Antibiotics 17, 110; in animal feed
43, 44
Antifreeze 69, 71, 72
Antioxidants 48, 50
Antiseptics 53
Arsenic 6, 26
ASA (see acetylsalicylic acid)
Asbestos 98, 100, 101
Aspergillus flavus 17
Asphyxiants 103